s o l u t i o n s @ s y n g r e s s . c o m

With more than 1,500,000 copies of our MCSE, MCSD, CompTIA, and Cisco study guides in print, we continue to look for ways we can better serve the information needs of our readers. One way we do that is by listening.

Readers like yourself have been telling us they want an Internet-based service that would extend and enhance the value of our books. Based on reader feedback and our own strategic plan, we have created a Web site that we hope will exceed your expectations.

Solutions@syngress.com is an interactive treasure trove of useful information focusing on our book topics and related technologies. The site offers the following features:

- One-year warranty against content obsolescence due to vendor product upgrades. You can access online updates for any affected chapters.
- "Ask the Author" customer query forms that enable you to post questions to our authors and editors.
- Exclusive monthly mailings in which our experts provide answers to reader queries and clear explanations of complex material.
- Regularly updated links to sites specially selected by our editors for readers desiring additional reliable information on key topics.

Best of all, the book you're now holding is your key to this amazing site. Just go to **www.syngress.com/solutions**, and keep this book handy when you register to verify your purchase.

Thank you for giving us the opportunity to serve your needs. And be sure to let us know if there's anything else we can do to help you get the maximum value from your investment. We're listening.

w w w . s y n g r e s s . c o m / s o l u t i o n s

SYNGRESS®

BUILDING ROBOTS
WITH LEGO®
MINDSTORMS

The ULTIMATE Tool for MINDSTORMS Maniacs!

Mario Ferrari
Giulio Ferrari
Ralph Hempel Technical Editor

KEY	SERIAL NUMBER
001	B8EL495GK4
002	2NVA4UHBBJ
003	CJGE946M43
004	3BVNAM7L4T
005	D384NSARSD
006	4ZMWAQEKFK
007	FMAPPW8GN9
008	XSLEKRK2FB
009	QMV9DSRUJT
010	5KNAPFRPAR

PUBLISHED BY
Syngress Publishing, Inc.
800 Hingham Street
Rockland, MA 02370

Building Robots with LEGO MINDSTORMS

Printed in the United States of America

3 4 5 6 7 8 9 0

ISBN: 1-928994-67-9

Technical Editor: Ralph Hempel
Co-Publisher: Richard Kristof
Acquisitions Editor: Catherine B. Nolan
Developmental Editor: Kate Glennon
Freelance Editorial Manager: Maribeth Corona-Evans

Cover Designer: Michael Kavish
Page Layout and Art by: Shannon Tozier
Copy Editor: Michael McGee
Indexer: Robert Saigh

Distributed by Publishers Group West in the United States and Jaguar Book Group in Canada.

Letter from the Publisher

When I co-founded Syngress in 1997 with Amorette Pedersen, we decided to forego the opportunity to include the ubiquitous "Letter from the Publisher" in the front of Syngress books. Our books are of the highest quality, written by content experts, and they've spoken quite well for themselves without any help from us.

However, the publication of *Building Robots with LEGO MINDSTORMS* entitles me to a one-time exemption from our rule. I am lucky enough to be the father of nine-year-old Sam Williams, who has taught me (among many important things) the joy of building with LEGO. Since helping Sam put together his first bricks at two years old to programming our latest MINDSTORMS robot (the optimistically named "Chore-Doer 3000"), I have derived hundreds of hours of pleasure creating projects with Sam. Perhaps the most ingenious thing about LEGO products, particularly the MINDSTORMS, is that the same product can be as challenging and enjoyable to a 43 year old as it is to a nine year old.

When presented with the chance to publish Mario and Giulio Ferrari's book, I jumped at the opportunity. As I read the manuscript, I could sense the authors had the same passion for creating with LEGO MINDSTORMS that Sam and I have. I knew immediately that there was a market of at least two people for the book!

I had the opportunity to meet Mario Ferrari at the Frankfurt Book Fair just weeks prior to this book's publication. I am American and Mario is Italian, but the language we spoke was that of two parents who have discovered a common passion to share with our nine and ten year old sons.

I hope you enjoy reading this book as much as we have enjoyed publishing it.

—*Chris Williams*
President, Syngress Publishing

Letters from the Authors

October 1998. It was a warm and sunny October and I remember it as if it was just yesterday. Giovanni, a colleague of mine, returns home to Italy from his honeymoon in New York. He carries in the office an enormous blue box whose cover reads "LEGO MINDSTORMS Robotics Invention System." When Giovanni opens the box and shows me the contents, I already know I must have one.

Let me go back to the late 70s. I was a high school student and had left my many years of LEGO play behind me. I was enthusiastically entering the rising personal computing era. Many of you are probably simply too young to remember that period, but "using" a computer mainly meant *programming* it. The computers of that time had few resources and rather primitive user interfaces; they were essentially mass storage devices, or something like a large unreliable cassette recorder. We programmers had to count and save every single byte, and even the most trivial tasks were very challenging. But at the same time, of course, it was great fun!

I developed a very strong interest in computer programming, and in Artificial Intelligence in particular. Machines and mechanical devices had always fascinated me, and it came quite naturally to me to turn to robotics as an expansion of this interest. There were some relatively cheap and compact computing devices that could provide a brain for my creatures, but unfortunately I discovered very soon all the technical problems involved in building the hardware of even a very simple robot. Where could I find motors? Which were the right ones? Where could I learn how to control them? What kind of gearing did I need? Imagine spending months folding aluminum plates, mounting bearings, assembling electronic circuits, connecting wires… and assuming you're able to do all those things, what do you get? A simple tin box that can run across the room and change direction when it hits an obstacle. The effort was definitely far greater than the results. Another problem was that constructing a new project meant starting again from scratch, with new materials. I wasn't patient enough, so I decided that a hobby in robotics was not for me.

The dream of robotics remained a dream. Until Giovanni opened *that* box. As soon as I got my hands on my first LEGO MINDSTORMS Robotics Invention System (RIS) set, it proved to be the fast and flexible robotics system that I was looking for. I found that the microcomputer, called the RCX, was very simple to use but powerful enough to let me drive complex devices. I became more and more intrigued by this toy, and through the Internet I soon discovered that I was not alone. It seemed an entire world of potential robotics fans had just been waiting for this product, and the LEGO company itself sold much more of them than expected.

From that October on, many things happened: I discovered LUGNET, the fantastic LEGO Users Group Network, the best resource ever for LEGO fans of any kind. I created a small Web site where I published pictures and information about my robotic creatures. Through these channels every day I got in touch with new people, and with some of them friendships have sprung up that go beyond our common interests in LEGO robotics. This is really the most special and valuable thing MINDSTORMS have given to me: Good friends all over the world.

—*Mario Ferrari*

October 1999. Another warm and sunny October, but on one particular day the Media Lab at the Massachusetts Institute of Technology (MIT) in Cambridge, MA has a different look. One large room at the facility is filled with exhibition tables with piles of colorful LEGO pieces and strange constructions on them and, there are hundreds of adults and children, LEGO bricks in hand, showing off their robotic creations and discussing the characteristics of their favorites. This is the world's biggest gathering of LEGO MINDSTORMS fans—the Mindfest!

When and how did all this start? It seems only yesterday to me, but a year had passed since I discovered MINDSTORMS for the first time. My brother Mario called me on the phone one evening, knowing I was about to leave on a short trip to New York, and asked me to bring him home a new product from LEGO, a sort of a programmable brick that could be controlled via a standard computer. I have to say that I was very curious, but nothing more: I thought it might be a great new toy to play around with, but I didn't completely understand its possibilities. When I saw the Robotics Invention System (RIS) in the toy store, though, I immediately realized how great it could be, and that I must have one, too. My own addiction to the LEGO MINDSTORMS began in that moment.

Like nearly everyone under the age of 40, I'd built projects from the many LEGO theme kits in my childhood. I had the advantage of using the large quantity of bricks that my older brothers and sisters had accumulated during the years, plus some new pieces and sets of the 80s. Castles, pirates, trains… hours and hours of pure fun, creating a large number of any kind of building and adventures. When I was a little older, I discovered the TECHNIC series, a wonderful world of machines, gears, mechanical tools, and vehicles, with endless construction possibilities. Then, like many other people, I abandoned LEGO as a young adult, and it remained out of my life—until I bought that big blue box in New York that day.

Why do I like LEGO MINDSTORMS so much? For me, it is mainly because it requires different skills and combines different disciplines: computer programming, robotics, and hands-on construction. You have to combine theory and practice, and to coordinate the design, construction, software, and testing processes. You can exercise your creativity and your imagination, and you have a great tool for doing this—a tool that is at the same time easy to use and very powerful, and most important, that doesn't limit your ideas.

And there's even more to the rewards of MINDSTORMS than that. Let's go back to Mindfest for a moment. Why would such an extraordinary group of people of different ages, cultures, and nationalities travel from all over the world to spend an entire weekend playing with LEGO? What exactly do they have in common? Why do some of the most famous Artificial Intelligence experts seriously discuss every feature of this product? There must be something really special about this "toy."

Joining an international community is one of the best things about playing with LEGO. It is not only a toy, but also a way of thinking and living. Just play with the MINDSTORMS for a while—you'll see for yourself!

—*Giulio Ferrari*

Author Acknowledgements

We would first like to thank Brian Bagnall for suggesting our names to Syngress Publishing when he heard Syngress was looking for an author to write a book focused on ideas and techniques for building MINDSTORMS robots. We are very grateful to Syngress Publishing for having turned this suggestion into a real opportunity, and for having allowed us the great freedom of deciding the shape and content of the book.

Additional appreciation goes to Jonathan Knudsen, who encouraged us in embarking upon the adventure of writing a book, and who helped us in understanding the world of technical publishing. Another friend, Guido Truffelli, patiently read every page of the manuscript. Many thanks, Guido—your comments and suggestions were very valuable in making the book more complete and more useful.

When Ralph Hempel accepted the offer to perform the technical edit of the book, we were really excited. Ralph's contributions to MINDSTORMS robotics are impressive, and range from mechanical solutions to extreme programming. His involvement proved to be even more significant that we had even imagined.

This was our first authoring experience, and all the Syngress staff has been incredibly patient with us and very supportive. A very special thank you goes to Kate Glennon, our Developmental Editor, for having taught us how to transform a collection of concepts and ideas into a book. Mario wants to also thank his employer, EDIS, which granted him the time to focus more attention on the book.

This book would have not been written without the contributions of the entire LUGNET MINDSTORMS Robotics community. Its members are incredibly creative, competent, helpful, and friendly, and they are always willing to share ideas and solutions with other people. We have attempted to give proper credit to all the people whose ideas we mentioned in the book, and we apologize in advance for those people who have been unintentionally left out.

Last but not least, we'd like to express enormous gratitude to our families, who encouraged and supported us through every moment of these intense months of writing.

Syngress Acknowledgements

We would like to acknowledge the following people for their kindness and support in making this book possible.

Richard Kristof and Duncan Anderson of Global Knowledge, for their generous access to the IT industry's best courses, instructors, and training facilities.

Karen Cross, Lance Tilford, Meaghan Cunningham, Kim Wylie, Harry Kirchner, Kevin Votel, Kent Anderson, and Frida Yara of Publishers Group West for sharing their incredible marketing experience and expertise.

Mary Ging, Caroline Hird, Simon Beale, Caroline Wheeler, Victoria Fuller, Jonathan Bunkell, and Klaus Beran of Harcourt International for making certain that our vision remains worldwide in scope.

Anneke Baeten and Annabel Dent of Harcourt Australia for all their help.

David Buckland, Wendi Wong, Daniel Loh, Marie Chieng, Lucy Chong, Leslie Lim, Audrey Gan, and Joseph Chan of Transquest Publishers for the enthusiasm with which they receive our books.

Kwon Sung June at Acorn Publishing for his support.

Ethan Atkin at Cranbury International for his help in expanding the Syngress program.

A special thanks to Sam Williams, who comes to the office every week with a backpack full of LEGOs. Watching the look on his face when he opens a new kit is a joyous event.

Contributors

Called the "DaVincis of LEGOs," Mario and Giulio Ferrari are world-renowned experts in the field of LEGO MINDSTORMS robotics.

Mario Ferrari received his first Lego box around 1964, when he was 4. Lego was his favorite toy for many years, until he thought he was too old to play with it. In 1998, the LEGO MINDSTORMS RIS set gave him reason to again have LEGO become his main addiction. Mario believes LEGO is the closest thing to the perfect toy and estimates he owns over 60,000 LEGO pieces. The advent of the MINDSTORMS product line represented for him the perfect opportunity to combine his interest in IT and robotics with his passion for LEGO bricks. Mario has been a very active member of the online MINDSTORMS community from the beginning and has pushed LEGO robotics to its limits. Mario is Managing Director at EDIS, a leader in finishing and packaging solutions and promotional packaging. He holds a bachelor's degree in Business Administration from the University of Turin and has always nourished a strong interest for physics, mathematics, and computer science. He is fluent in many programming languages and his background includes positions as an IT manager and as a project supervisor. Mario works in Modena, Italy, where he lives with his wife Anna and his children Sebastiano and Camilla.

Giulio Ferrari is a student in Economics at the University of Modena and Reggio Emilia, where he also studied Engineering. He is fond of computers and has developed utilities, entertainment software, and Web applications for several companies. Giulio discovered robotics in 1998, with the arrival of MINDSTORMS, and held an important place in the creation of the Italian LEGO community. He shares a love for LEGO bricks with his oldest brother Mario, and a strong curiosity for the physical and mathematical sciences. Giulio also has a collection of 1200 dice, including odd-faced dice and game dice. He studies, works, and lives in Modena, Italy.

Technical Editor

Ralph Hempel (BASc.EE, P.Eng) is an Independent Embedded Systems Consultant. He provides systems design services, training, and programming to clients across North America. His specialty is in deeply embedded microcontroller applications, which include alarm systems, automotive controls, and the LEGO RCX system. Ralph provides training and mentoring for software development teams that are new to embedded systems and need an in-depth review of the unique requirements of this type of programming. Ralph holds a degree in Electrical Engineering from the University of Waterloo and is a member of the Ontario Society of Professional Engineers. He lives in Owen Sound, Ontario with his family, Christine, Owen, Eric, and Graham.

Contents

Learn about Lego Gears

Explore LEGO Sensors

LEGO sensors come in two families: *active* and *passive* sensors. Passive simply means they don't require any electric supply to work. The touch and temperature sensors belong to the passive class, while the light and rotation sensors are members of the active class.

Understand the Benefits of Designing Modular Code

- Readability
- Reusability
- Testability

Create Custom Components

Explore extra parts, custom sensors, and tricks for using the same motor for more than one task:

- Extra parts come from either sets or service packs.

- Custom sensors are a new frontier, and reveal a whole new world of possibilities.

- Mechanical tricks enable you to use the same motor to power multiple mechanisms.

Use Ankle Bending Techniques

Use Angle Connectors

There are currently six types of angle connectors in the LEGO line, numbered 1 to 6. In case you're wondering how the numbers relate to angles, here are the correspondences: 1 = 0°, 2 = 180°, 3 = 157.5°, 4 = 135°, 5 = 112.5°, 6 = 90°. They go by increments of 22.5°, a quarter of a right angle.

Build a Pianist

This robot requires a lot of extra parts, mainly beams and plates used to make the structure solid enough to withstand the forces involved in the performance.

Understand Infrared Communication

Infrared (IR) light is of the same nature as visible light, but its frequency is below that perceivable by the human eye. Provided the intensity is high enough, we usually feel IR radiation as heat.

Design Other Useful Robots

- Alarm Clock
- Baby Entertainer
- Pet Feeder
- Dog Trainer

Find Useful Lego Sites

- www.brickshelf.com
- http://fredm.www
  .media.mit.edu/people/
  fredm/mindstorms/
  index.html
- www.crynwr.com/
  lego-robotics/
- www.bvandam.net

Foreword

Like many other programmers, I credit my early years of playing with LEGO as a major factor in my future career path. As my family and I watched the United States launching the Apollo 11 rocket, I was playing with a LEGO truck—it was my birthday and I was 7 years old. What I could not know at the time was that 30 years later I would hold in the palm of my hand a microcontroller with more raw speed and memory than the one the astronauts used to get to the moon and back. That computer would be encased in yellow ABS plastic and would change the world of hobby and educational robotics.

The story of my involvement with the LEGO MINDSTORMS is a familiar one. Discussion of building a custom controller for LEGO TECHNIC creations was a frequent topic in Lugnet (the LEGO Users Group) discussion forums. I had doubts about our ability to make a controller that everyone could afford. Then LEGO released MINDSTORMS in the fall of 1988—and I just had to have one.

Within weeks of the release, Kekoa Proudfoot had "cracked" the protocol between the RCX brick and the desktop computer, and he soon had a complete disassembly of the object code online. Using this as a base, intrepid programmers like Marcus Noga and Dave Baum soon had alternative programming environments for the RCX—including my own contribution, called pbForth. On the hardware front, Michael Gasperi figured out how the sensor and motor ports worked and contributed his knowledge freely.

LEGO had an unbelievable hit on their hands. The sales of the MINDSTORMS kits exceeded their wildest predictions, and more than half the sales were to adults! When the Massachusetts Institute of Technology (MIT) asked me to participate in a panel at the Mindfest gathering in 1999, I was honored to be there with the likes of Dave Baum, Michael Gasperi, Marcus Noga, and Kekoa Proudfoot. In our panel discussion, we discussed how the Internet had made it possible for widely separated people to work together.

While at Mindfest, I met Mario and Giulio Ferrari. They had their Tic Tac Toe robot set up for demonstrations and it was a big hit. The brothers immediately struck me as energetic and dedicated LEGO hobbyists. The other members of the Italian ITLug group have provided LUGNET readers a steady stream of wonderful robots in the past few years.

I have had the pleasure of watching children and adults of all ages build machines and robots with their MINDSTORMS kits. In almost all cases their initial attempts ended in frustration with their *mechanical* skills. In fact, many builders never even get to the stage of programming their robots. This book will be a welcome addition to their libraries because of the vast amount of information it contains. From basic bracing techniques to drive and grip mechanisms—it's all here. Even if a particular robot does not appeal to the reader, the ideas used in its construction may be transferred to other robots in unusual and surprising ways.

As a co-author of *Extreme Mindstorms*, a book about programming the RCX, I appreciate the effort that went into this book. Mario and Giulio have taken the time to guide the reader through the basics of building their creations by setting realistic performance goals and then experimenting with different methods. This important skill goes by the unassuming name of *tinkering*, and cannot be underestimated. The MINDSTORMS system gives the hobby and educational market a modular and inexpensive way to develop these important tinkering skills.

As the technical editor of this volume, I have had my own creativity sparked by some of the robots Mario and Giulio have documented. I am amazed at the sheer volume of ideas, the quality of the photos, and the careful presentation of ideas that many readers will encounter for the first time. The staff at Syngress Publishing has been a pleasure to work with, and they deserve credit for bringing the hard work of the Ferrari brothers to the wide audience that I'm sure this book will enjoy.

So clear some space on a table, open this book and get out your MIND-STORMS set, and start tinkering!

—*Ralph Hempel*

Preface

Why Robotics?

What's so special about robotics? Why have LEGO MINDSTORMS experienced such great success? Each one of us might have our own answers.

Robotics is an interdisciplinary subject, combining different fields of study that in traditional educational systems you usually examine separately: physics, mathematics, electronics, and computer programming, just to name a few. Robotics is a hobby through which you can find a practical application for many of the concepts you studied in school—or, if you didn't study them, or don't have an aptitude for them, it offers a great way to learn by experience and by having some fun. The most important point, however, is that robotics is *more* than the sum of the basic notions you're required to know. It gives you a precise and concrete idea of how these notions integrate and complete each other. So it happens that when you're looking for a solution to a problem, by following your intuition and knowledge it's almost a given that you'll find a solution different from that devised by someone else.

Let's say you have just built your first line-following robot (we'll discuss this topic in detail in Part II). You discover that your robot works, but it makes too many corrections to its steering and this affects its resulting speed. What could you do to fix it? If you have a talent for mechanics, your first approach might be to try and modify the structure and architecture of your robot. You might observe that the wheels are too close to each other in your differential drive, and for this reason your robot turns very fast and tends to over-correct its steering. Or you might decide that the differential drive architecture after all is not the best option for line following. You may even discover that the position of the light sensor in the robot greatly affects its performance.

If you are an experienced programmer, you might instead work out your code to correct the robot's behavior. You feel at ease with timers and counters, so you change the program to introduce some delay in the route changes, then you spend some in time in testing and trimming it until you find an optimum value for the constants you used.

At the same time, if you have a decent understanding of physics, you could reach into your knowledge base for something useful, and discover a model you were taught when studying magnetism: hysteresis (if you don't know what hysteresis is, don't worry, we'll explain it in Part I!). You realize that you can make your robot follow a different scheme when going from black to white rather then from white to black. You think that this might improve its performance—and it actually does.

What lesson should we learn from this example? That there's no one unique solution, there are many of them. And the more you are able to open your mind and explore new possible approaches, the higher your chances of working out a solution. Robotics does not involve a list of techniques to follow in order, rather it is a process in which your creativity plays a very strong role, allowing you to follow a new path to the goal each time.

There's another element that makes robotics so interesting to us and, I suspect, to many other people as well. It forces you to look at the world with new eyes, those of a child's.

If you observe babies exploring the environment, you will notice that they are surprised by everything. They don't take anything for granted. They try everything, continuously developing new concepts by testing new approaches. We adults usually laugh at most of those attempts, to our mature minds they seem absurd, either because we already know that a specific thing is impossible to do, or because we know the solution to the problem the child is tackling. When approaching tasks in robotics, we are forced to become children again, to rediscover the world with different senses.

Let's look at this concept using another example: You are new to robotics, facing your first project, but are wise enough to decide on a very simple task. You want to create a robot that's able to move around your house. You naturally want your robot to be able to detect obstacles when it hits them, so it can change direction and toddle off on a new path. You design your mechanical marvel so it can go forward, backward, and change direction. Then you add a simple bumper to detect obstacles, something that closes a switch when pressed. Finally, you write some code so your robot is ready for its debut on the living room floor—but wait, you forgot about the shag carpet, and carpet loops get into your gears and mess everything up. You decide testing might be better in the kitchen. Now your robot runs well; it hits a wall, turns on itself, and spins off in another direction. Up to this point, it's a pleasure to watch…but then it runs up against a sideboard, and the upper part of the robot gets blocked by the furniture, preventing the lower bumper from detecting the obstacle. Okay, so you have to improve the bumper. In the meantime, you break down and manually turn the robot in a new direction. Hey! Pay attention! It's heading to the basement stairs! Rescue it and add edge detection to your list of improvements. You will learn quickly that even a simple action like climbing the stairs is the result of a very, very complex balancing of weights and strengths, precise positioning, and coordination.

If you have kids in your circle of family and friends, you will have the precious opportunity to watch how they interact with your robots. In our experience, young kids tend to expect a lot from robots—a lot more than what simple inexpensive robots can currently do. They have forgotten all the difficulties they had to overcome themselves, and they're still naïve enough to believe that all the amazing things robots do in movies can be carried out by your robots as well. They see any possible task or function as easy to implement. "Why don't you make one like the *real* R2-D2, Dad?"

MINDSTORMS provides a great way for kids to understand that even the most common activities are composed of many individual operations. If they don't understand, if they become frustrated by what the robots *can't* do, play an easy but funny game with them in which *you* are the robot and they have to "program" you using only a very simple vocabulary describing a few basic actions. They will laugh at all the stupid things you'll do and the unusual situations their commands will get you into—but they will very likely understand the point. This is an extra gift that robotics will provide to your family: showing your children how to deconstruct and analyze what they consider a single action.

When you're really at a loss for what robot you might build next, ask the kids! You're sure to get a bunch of fresh ideas. Most of us tend to design robots that move around, grab objects, find soda cans in a room or do any other activity we *expect* robots should do. Some of these projects are very challenging, and most are very instructive. But if you ask the kids what they would like to see, you get responses like: "Why don't we build a *skiing* robot, Dad?" Would you ever think of a skiing robot? Just the same, robots of this type are easy to make (see Chapter 16). They require only basic parts, they're fun, and like any MINDSTORMS challenge, they're definitely worth the time you spend on them.

Why LEGO?

If you've been raised with LEGO like we have, you already know what's special about it. But for those relatively new to the LEGO concept, including those who have yet to buy a MINDSTORMS set, let us explain why LEGO is an excellent choice for exploring the world of robotics.

The power of the LEGO system lies in its founding concept: reusability. The same basic brick can today be the foot of an elephant, tomorrow a block in an Egyptian pyramid, and the day after the nose of a robot. When you open a LEGO box, you see the parts that will form a LEGO model, but you also see an infinite number of possible models you might create with those parts.

The property that transforms these small plastic pieces into a construction system is their connectivity. You don't need glue, screws, or any special tools (other then your hands) to assemble (or dismantle) a LEGO model. The LEGO parts easily snap on to each other and stay firmly in place until you decide to take them apart. The parts won't be damaged, no matter how much you use them.

But what really makes LEGO easy to use is its modularity. Not only does one brick connect to another, but they do so at predefined, discrete positions. There are studs and holes that force you to assemble parts following a precise geometric scheme. This might seem a limitation at first, but it actually makes your life easier because of something called precise positioning. You don't need a ruler or a square—all that's required is that you can count!

- **LEGO is fast** You don't have to saw, cut, drill, solder, fold, file, or mill your components. They are ready to use, just pick up what you need from the box.

- **LEGO is clean** You don't produce filings, don't need any lubricants or paints, and when you have finished playing with it, your room looks exactly as it did before. This is a very important point to make to the people who live with you if you want them to be tolerant of your hobby!

- **LEGO is cost-effective** You can use and reuse your LEGO parts as needed to produce many generations of robots. And should you ever eventually tire of your LEGO pieces, they will still have a market value. There are other easy-to-assemble robotic kits on the market, but they usually only permit you to build one specific model. Beyond that, there's nothing more you can get from their kits.

- **LEGO is ecological** We don't mean that its ABS plastic is easy to recycle. It is, but that's not the point. You simply don't need to recycle it, because you'll never throw it away. After all, this is the most respectful approach to the environment: making products with a long life span, that don't exhaust their function and don't require recycling or disposal. We still use many of the LEGO bricks we received during childhood.

To return to robotics, some of you may believe that LEGO MINDSTORMS is too limited a system to build sophisticated projects. This is true if you mean *really* sophisticated systems! Others may observe that LEGO is not suitable for building robots that perform actual work. This, again, can be true, although we will show some examples in Chapter 25 that are indeed useful work projects. The LEGO MINDSTORMS kit is definitely more than a toy—it is probably the most fun and effective educational tool for learning the scientific principles behind robotics. There are indeed limitations, but this is part of the fun, challenging you to use your imagination, to find esoteric solutions for seemingly unsolvable problems.

Suppose you're an experienced programmer, asked to write the umpteenth version of an invoicing software—just the thought of it puts you to sleep. But then your employer adds "Oh, by the way, it has to run on a machine with 3 K of RAM. *Now* you're interested! After all, there's nothing like a challenge.

So, don't feel limited by the constraints you find in the system, feel inspired. Create a robot that makes your friends say "I didn't think it was possible to make such an incredible thing with LEGO!" Because you can.

Using this Book

This book is about building robots using LEGO bricks and components. The chapters in Part I are about *how* to build a robot. Here, we provide a set of *tools* you'll need to explore the world of robotics. We'll review basic knowledge about mechanics, motors, sensors, pneumatics, and navigation. We will compare different standard architectures, discuss solutions to common recurring problems, and will suggest how to organize complex projects in terms of subsystems.

Part II will face the tough question, "I've got my MINDSTORMS kit, I've learned how to use it—so *what* do I build?" Here we will show you a large survey of possible ideas, but do not expect to find complete models to build step by step. The goal of this book is *not* to teach you to re-create our models, instead it is to stimulate your imagination to create your own. Imagination and creativity cannot be *taught*, but it can be inspired. We hope that our approach might help you see the world with different eyes. The same is true for understanding the mechanics of robotics: you will learn best by guided and informed experimentation. Actively participating in the process, not simply cloning our models, will bring you the greatest rewards.

Part III takes you into the world of robotics contests. These contests offer LEGO builders a challenge beyond the initial goal of building a working robot—they provide a means to inspire ideas, share solutions, and just have fun, whether with your own friends, in a local group, or even internationally. There are different ways to attend a robotic contest: you can compare robots with friends in person, or you can take up a challenge someone has organized through the Internet, in which case you submit your solution in pictures or programming code. Either way, you will learn a great deal from your opponents. And from the rules, too: what really makes a contest exciting is trying to find an original but "legal" solution you hope your opponents haven't thought of.

The last part of the book consists of appendices that provide you with various technical resources we hope will be helpful to you.

There is a key element to robotics that you will *not* find in this book: comprehensive coverage of programming and electronics. We made a conscious choice to focus this book on construction solutions and to cover only as much programming as was necessary—a limited amount of coverage is indeed required, because you cannot successfully design and build your robots without taking into consideration the role that electronics will play. Because there are various programming options you can choose from, depending on your level of programming experience, we have written our code using NQC, a very widespread C-like textual language that you can easily translate into your favorite language.

One of the nicest things about MINDSTORMS robots is that you're not required to be an electrical engineer to design them—we're not! If you are interested in expanding your RCX possibilities on that side, we will point you to the right resources.

Please note that we don't expect you read the book sequentially from cover to cover: feel free to jump to a specific page or topic. When we cite a concept or technique explained in a previous chapter, we'll tell you where to find it. The only things we expect from you are the following:

- That you own a MINDSTORMS Robotic Invention System kit or you are seriously interested in buying one. Many of the tips and ideas are, however, applicable to other LEGO programmable bricks (such as Scout and Cybermaster) or to nonrobotic LEGO TECHNIC models.

- That you already have some basic skill in assembling LEGO TECHNIC parts and in programming your RCX. Doing the lessons included in the MINDSTORMS CD-ROM and being familiar with the Constructopedia will be all the background you need.

Enjoy our book!

Part I

Tools

Chapter 1

Understanding LEGO® Geometry

Solutions in this chapter:

- Expressing Sizes and Units
- Squaring the LEGO World: Vertical Bracing
- Tilting the LEGO World: Diagonal Bracing
- Expressing Horizontal Sizes and Units
- Bracing with Hinges

Introduction

Before you enter the world of LEGO® robotics, there are some basic geometric properties of the LEGO bricks we want to be sure you know and understand. Don't worry, we're not going to test you with complex equations or trigonometry, we'll just discuss some very simple concepts and explain some terminology that will make assembling actual systems easier from the very beginning.

You will discover which units LEGO builders use to express sizes, what the proportions of the bricks are, and how this affects the way you can combine bricks with different orientations into a solid structure.

We encourage you to try and reproduce all the examples we show in this chapter with your own LEGO parts. Keep your MINDSTORMS box handy so you can pick up the parts you need, which in this chapter will actually be nothing more than a few bricks and plates.

If, for any reason, you feel the stuff here is too complex or boring, don't force yourself to read it, skip the chapter and go to another one. You can always come back and use this chapter as a sort of glossary whenever it's needed.

Expressing Sizes and Units

LEGO builders usually express the size of LEGO parts with three numbers: *width*, *length*, and *height*, in that order. The standard way to use LEGO bricks is "studs up." When expressing sizes, we always refer to this orientation, even when we are using the bricks upside down or when rotating them in 3-D space.

Height is the simplest property to identify, its the vertical distance between the top and bottom of the basic brick. Width, by convention, is the shorter of the two dimensions that lie on the horizontal plane (length is the other one). Both width and length are expressed in terms of *studs*, also called *LEGO units*. Knowing this, we can describe the measurements of the most traditional brick, the one whose first appearance dates back to 1949, which is 2 x 4 x 1 (see Figure 1.1).

LEGO bricks, although their measurements are not expressed as such, are based on the metric system—a stud's width corresponds to 8mm and the height of a brick (minus the stud) to 9.6mm. These figures are not important to remember—what's important is that they do not have equal values, meaning you need two different units to refer to length and height. Their *ratio* is even more important: dividing 9.6 by 8 you get 1.2 (the vertical unit corresponds to 1.2 times the horizontal one). This ratio is easier to remember if stated as a proportion between whole numbers: It is equivalent to 6:5. We will explore the relevance of this ratio in the next section.

Figure 1.1 Measurements of a Traditional LEGO Brick

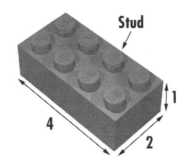

Figure 1.2 shows the smallest LEGO brick, described in LEGO units as a 1 x 1 x 1. For the reasons explained previously this LEGO "cube" is not a cube at all.

Figure 1.2 Proportions in a 1 x 1 x 1 LEGO Brick

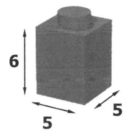

The LEGO system includes a class of components whose height is one-third of a brick. The most important element of this class is the *plate*, which comes in a huge variety of rectangular sizes and in some special shapes, too. If you stack three plates, you get the height of a standard brick (see Figure 1.3).

Figure 1.3 Three Plates Make One Brick in Height

Squaring the LEGO World: Vertical Bracing

Why do we care about all these relationships? To answer this, we must travel back to the late seventies when the LEGO TECHNIC line was created. Up to that time, LEGO was designed and used to build things made of horizontal layers: Bricks and plates integrate pretty well when stacked together. Every child soon learns that three plates count for a brick, and this is all they need to know. But in 1977, LEGO decided to introduce a new line of products targeting an older audience: LEGO TECHNIC. They gave the common 1xN brick holes and turned it into what we call a TECHNIC brick, or a *beam* (Figure 1.4). These holes allow *axles* to pass through them, and also permit the beams to be connected to each other via *pegs*, thus creating an entire new world of possibilities.

Figure 1.4 The LEGO TECHNIC Beam

Suppose you want to mount a beam in a vertical position, to brace two or more layers of horizontal beams. Here's where we must remember the 6 to 5 ratio. The holes inside a beam are spaced at exactly the same distance as the studs, but are shifted over by half a stud. So, when we stand the beams up, the holes follow the horizontal units and not the vertical ones. Consequently, they don't match the corresponding holes of the layered beams. In other words, the holes in the vertical beam cannot line up with the holes in the stack because of the 6:5 ratio. At least not with all the holes. But let's take a closer look at what happens. Count the vertical units by multiples of 6 (6, 12, 18, 24, 30…) and the horizontal ones by multiples of 5 (5, 10, 15, 20, 25, 30…). Don't count the starting brick and the starting hole, they are your reference point; you are measuring the *distances* from that point. You see? After counting 5 vertical units you reach 30, which is the same number you reach after counting 6 horizontal units (see Figure 1.5).

Is there a general rule we can derive from this? A sort of theorem? Yes: *In a stack of horizontal beams, at every fifth beam the holes align to those of a perpendicular beam.*

Figure 1.5 Matching Horizontal and Vertical Beams

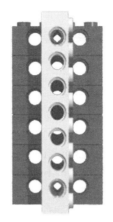

Now you can build a stack with some of your beams, brace them with another long one, and verify this rule in practice. If you put an axle in the first connecting hole and then try to put it again in the following holes, you'll find that the holes of the crossed beam match at the starting brick plus five and at the starting brick plus ten (see Figure 1.6).

This technique of crossing beams is extremely important. It's what enables us to build solid models, because the vertical beam locks all the beams in between the two horizontal beams. It's a pity we need to stack 6 beams before we can lock them with a traverse beam. Couldn't we build something more compact? The answer is, of course, yes.

Recall that the vertical unit has a subunit, the height of a plate. Three plates make a brick, so counting plates, we can increase the height by steps of 2 instead of 6 (2 is one-third of 6). Our progression in height now becomes: 2, 4, 6, 8, 10… after 5 vertical increments we reach the value 10. That's also in the horizontal scale of values, a spot where we know the holes will match. So our new and final theorem is: *every 5 plates in height, the holes of perpendicular beams match*. If there's a single thing you should remember from this chapter, this is it.

Unfortunately a plate cannot be used *as is* to connect a vertical beam, for the simple reason it hasn't any holes! But a beam is equivalent to three plates in height. Knowing this, we can state our rule in operational terms: Starting from the beam at the bottom (don't count it), add 1 for each plate and 3 for each beam, and keep at least a beam at the top. If the result is a multiple of 5, the holes can be matched by a perpendicular beam.

Figure 1.6 Every Five Bricks in Height the Holes Match

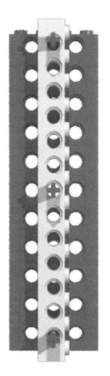

The most compact scheme that allows you to lock your horizontal layers with a vertical beam is the one shown in Figure 1.7: a beam and two plates, corresponding to five plates. Two holes per five plates is the only way you can connect bracing beams at this distance. You can find it recurring in all TECHNIC models designed by LEGO engineers, and we will use it extensively in the robots in this book.

Figure 1.7 The Most Compact Locking Scheme

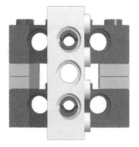

Upon increasing the distances, the possibilities increase; the next working combination is 10 plates/4 holes. But there are many ways we can combine beams and plates to count 10 plates in height; you can see some of them in Figure 1.8.

Figure 1.8 The Standard Grid

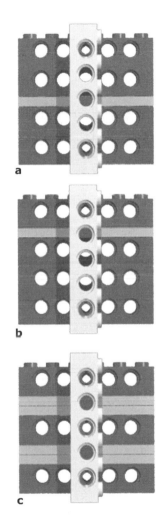

First question: Is there a best grid, a preferred one? Yes, there is, in a certain sense. The most versatile is version c in Figure 1.8, which is a multiple of our basic scheme from Figure 1.7, because it lets you lock the beams in an intermediate point, also. So, when you build your models, the sequence 1 beam + 2 plates + 1 beam + 2 plates... is the one that makes your life easier: Connections are possible at every second hole of the vertical beam. This is what Eric Brok on his Web site calls a *standard grid* (see Appendix A), a grid that maximizes your connection possibilities. Second question: Should you always stay with this scheme? Absolutely not! Don't curb your imagination with unnecessary constraints. This is just a tip that's useful in many circumstances, especially when you start something and don't know yet what exactly you're going to get! In many, many cases we use different schemes, and the same will be true for you.

Tilting the LEGO World: Diagonal Bracing

Who said that the LEGO beams *must* connect at a right angle to each other? The very nature of LEGO is to produce squared things, but diagonal connections are possible as well, making our world a bit more varied and interesting, and giving us another tool for problem solving.

You now know that you can cross-connect a stack of plates and beams with another beam. And you know how it works in numerical terms. So how would you brace a stack of beams with a diagonal beam?

You must look at that diagonal beam as if it was the hypotenuse of a right-angled triangle. Look at or build a stack like that in Figure 1.9. Now proceed to measure its sides, remembering not to count the first holes, because we measure lengths in terms of distances from them. The base of the triangle is 6 holes. Its height is 8 holes: Remember that in a standardized grid every horizontal beam is at a distance of two holes from those immediately below and above (we placed a vertical beam in the same picture to help you count the holes). In regards to the hypotenuse, it counts 10 holes in length.

For those of you who have never been introduced to Pythagoras, the ancient Greek philosopher and mathematician, the time has come to meet him. In what is probably the most famous theorem of all time, Pythagoras demonstrated that there's a mathematical relationship between the length of the sides of right-angled triangles. The sides composing the right angle are the catheti—let's call them A and B. The diagonal is the hypotenuse—let's call that C. The relationship is:

$A^2 + B^2 = C^2$

Now we can test it with our numbers:

$6^2 + 8^2 = 10^2$

This expands to:

(6 x 6) + (8 x 8) = (10 x 10)

36 + 64 = 100

100 = 100

Figure 1.9 Pythagoras' Theorem

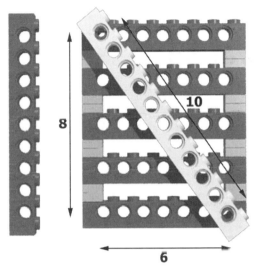

Yes! This is exactly why our example works so well. It's not by chance, it's good old Pythagoras' theorem. Reversing the concept, you might calculate if any arbitrary pair of base and height values brings you to a working diagonal. This is true only when the sum of the two lengths, each squared, gives a number that's the perfect square of a whole number. Let's try some examples (Table 1.1).

Table 1.1 Verifying Working Diagonal Lengths

A (Base)	B (Height)	A^2	B^2	$A^2 + B^2$	Comments
5	6	25	36	61	This doesn't work.
3	8	9	64	73	This doesn't work.
3	4	9	16	25	This works! 25 is 5 x 5.
15	8	225	64	289	This works too, though 289 is 17 x 17, this would come out a very large triangle.

Continued

Table 1.1 Continued

A (Base)	B (Height)	A²	B²	A² + B²	Comments
9	8	81	64	145	145 is not the square of a whole number, but it is so close to 144 (12 x 12) that if you try and make it your diagonal beam it will fit with no effort at all. After all, the difference in length is less than 1 percent.

At this point, you're probably wondering if you have to keep your pocket calculator on your desk when playing with LEGO blocks, and maybe dig up your old high school math textbook to reread. Don't worry, you won't need either, for many reasons:

- First, you won't need to use diagonal beams very often.

- Most of the useful combinations derive from the basic triad 3-4-5 (see the third line in Table 1.1). If you multiply each side of the triangle by a whole number, you still get a valid triad. By 2: 6-8-10 (the one of our first example), by 3: 9-12-15, and so on. These are by far the most useful combinations, and are very easy to remember.

- We provide a table in Appendix B with many valid side lengths, including some that are not *perfect* but so close to the right number that they will work very well without causing any damage to your bricks.

We suggest you take some time to play with triangles, experimenting with connections using various angles and evaluating their rigidity. This knowledge will prove precious when you start building complex structures.

Expressing Horizontal Sizes and Units

So far we've put a lot of attention into the vertical plane, because this technique of layers locked by vertical beams is the most important tool you have to build rock solid models. Well, almost rock solid, considering it's just plastic!

Nevertheless there are some other ideas you'll find useful when using bricks in the horizontal plane, that is, all studs up.

We said that the unit of measurement for length is the *stud*, meaning that we measure the length of a beam counting the number of studs it has. The holes in the beams are spaced at the same distance, so we can equally say "a length of three studs" or "a length of three holes." But looking at your beams, you have probably already noticed that the holes are interleaved with the studs, and that there is one hole less then the number of studs in each beam.

There are two important exceptions to this rule: the 1 x 1 beam with one hole, and the 1 x 2 beam with two holes (Figure 1.10). You won't find any of them in your MINDSTORMS box, but they're so useful you'll likely need some sooner or later.

Figure 1.10 The 1 x 1 Beam with 1 Hole and the 1 x 2 Beam with 2 Holes

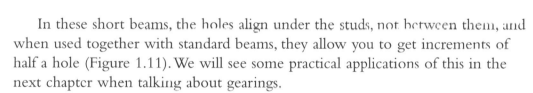

In these short beams, the holes align under the studs, not between them, and when used together with standard beams, they allow you to get increments of half a hole (Figure 1.11). We will see some practical applications of this in the next chapter when talking about gearings.

Figure 1.11 How to Get a Distance of Half a Hole

Another piece that carries out the same function is the 1 x 2 plate with one stud. This one also is not included in your MINDSTORMS kit, but it's definitely a very easy piece to find. As you can see in Figure 1.12, it's useful when you want to adjust by a distance of half a stud, and can help you a lot when fine tuning the position of touch sensors in your model. We'll see some examples of usage later on in this book.

Figure 1.12 The Single Stud 1 x 2 Plate

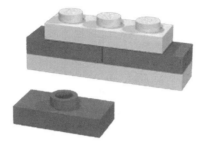

Bracing with Hinges

To close the chapter, we return to triangles. Before you start to panic, just think—you already have all the tools you need to manage them painlessly. There's nothing actually new here, just a different application of the previous concepts. Let us say in addition, that it's a technique you can survive without. But for the sake of completeness, we want to introduce it also.

First of all we need yet another special part, a *hinge* (Figure 1.13). Using these hinges you can build many different triangles, but once again our interest is on right-angle triangles, because they are by far the most useful triangle for connections. Their catheti align properly with lower or upper layers of plates or beams, offering many possibilities of integration with other structures.

Figure 1.13 The LEGO Hinge

The LEGO hinges let you rotate the connected beams, keeping their inner corners always in contact. Therefore, using three hinges, you get a triangle whose vertices fall in the rotation centers of the hinges. The length of its *inner* sides is the length of the beams you count (Figure 1.14). Regarding right-angled triangles: You're already familiar with the Pythagorean Theorem, and it applies to this case as well. The same combinations we have already seen work in this case: 3-4-5, 6-8-10, and so on.

Figure 1.14 Making a Triangle with Hinges

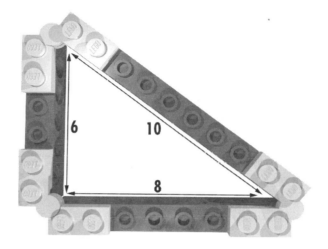

Summary

Did you survive the geometry? You can see it doesn't have to be that hard once you get familiar with the basics. First, it helps to know how to identify the bricks by their proportions, counting the length and width by studs, and recognizing that the vertical unit to horizontal unit ratio is 6 to 5. Thus, according to the simple ratio, when you're trying to find a locking scheme to insert axles or pins into perpendicular beam holes, you know that every 5 bricks in height, the holes of a crossed beam match up. Also, because three plates match the height of a brick, the most compact locking scheme is to use increments of two plates and a brick, because it gives you that magic multiple of 5. If you stay with this scheme, the standard grid, everything will come easy: one brick, two plates, one brick, two plates...

To fit a diagonal beam, use the Pythagorean Theorem. Combinations based on the triad of 3-4-5 constitute a class of easy-to-remember distances for the beam to make a right triangle, but there are many others. Either use the rules explained here, or simply look up the connection table provided in Appendix B.

Chapter 2

Playing with Gears

Solutions in this chapter:

- **Counting Teeth**
- **Gearing Up and Down**
- **Riding That Train: The Geartrain**
- **Worming Your Way: The Worm Gear**
- **Limiting Strength with the Clutch Gear**
- **Placing and Fitting Gears**
- **Using Pulleys, Belts, and Chains**
- **Making a Difference: The Differential**

Introduction

You might find yourself asking: Do I really *need* gears? Well, the answer is yes, you do. Gears are so important for machines that they are almost their symbol: Just the sight of a gear makes you think *machinery*. In this chapter, you will enter the amazing world of gears and discover the powerful qualities they offer, transforming one force into another almost magically. We'll guide you through some new concepts—velocity, force, torque, friction—as well as some simple math to lay the foundations that will give you the most from the machinery. The concepts are not as complex as you might think. For instance, the chapter will help you see the parallels between gears and simple levers.

We invite you once again to experiment with the real things. Prepare some gears, beams, and axles to replicate the simple setups of this chapter. No description or explanation can replace what you learn through hands-on experience.

Counting Teeth

A single gear wheel alone is not very useful—in fact, it is not useful at all, unless you have in mind a different usage from what it was conceived for! So, for a meaningful discussion, we need at least two gears. In Figure 2.1, you can see two very common LEGO gears: The left one is an 8t, while the right is a 24t. The most important property of a gear, as we'll explain shortly, is its *teeth*. Gears are classified by the number of teeth they have; the description of which is then shortened to form their name. For instance, a gear with 24 teeth becomes "a 24t gear."

Figure 2.1 An 8t and a 24t Gear

Let's go back to our example. We have two gears, an 8t and a 24t, each mounted on an axle. The two axles fit inside holes in a beam at a distance of two holes (one empty hole in between). Now, hold the beam in one hand, and with the other hand gently turn one of the axles. The first thing you should notice is that when you turn one axle, the other turns too. The gears are *transferring motion from one axle to the other*. This is their fundamental property, their very nature. The second important thing you should notice is that you are not required to apply much strength to make them turn. Their teeth match well and there is only a small amount of friction. This is one of the great characteristics of the LEGO TECHNIC system: Parts are designed to match properly at standard distances. A third item of note is that the two axles turn in opposite directions: one clockwise and the other counterclockwise.

A fourth, and more subtle, property you should have picked up on is that the two axles revolve at different speeds. When you turn the 8t, the 24t turns more slowly, while turning the 24t makes the 8t turn faster. Lets explore this in more detail.

Gearing Up and Down

Let's start turning the larger gear in our example. It has 24 teeth, each one meshing perfectly between two teeth of the 8t gear. While turning the 24t, every time a new tooth takes the place of the previous one in the contact area of the gears, the 8t gear turns exactly one tooth, too. The key point here is that you need to advance only 8 teeth of the 24 to make the small gear do a complete turn (360°). After 8 teeth more of your 24, the small gear has made a second revolution. With the last 8 teeth of your 24, the 8t gear makes its third turn. This is why there is a difference in speed: For every turn of the 24t, the 8t makes three turns! We express this relationship with a ratio that contains the number of teeth in both gears: 24 to 8. We can simplify it, dividing the two terms by the smaller of the two (8), so we get 3 to 1. This makes it very clear, in numerical terms, that one turn of the first corresponds to three turns of the second.

You have just found a way to get more speed! (To be technically precise, we should call it *angular velocity*, not *speed*, but you get the idea). Before you start imagining mammoth gear ratios for racecar robots, sorry to disappoint you—there is no free lunch in mechanics, you have to pay for this gained speed. You pay for it with a decrease in *torque*, or, to keep in simple terms, a decrease in strength.

So, our gearing is able to convert torque to velocity—the more velocity we want, the more torque we must sacrifice. The ratio is exactly the same, if you get three times your original angular velocity, you reduce the resulting torque to one third.

One of the nice properties of gears is that this conversion is symmetrical: You can convert torque into velocity or vice versa. And the math you need to manage and understand the process is as simple as doing one division. Along common conventions, we say that we *gear up* when our system increases velocity and reduces torque, and that we *gear down* when it reduces velocity and increases torque. We usually write the ratio 3:1 for the former and 1:3 for the latter.

Bricks & Chips...

What Is Torque?

When you turn a nut on a bolt using a wrench, you are producing *torque*. When the nut offers some resistance, you've probably discovered that the more the distance from the nut you hold the wrench, the less the force you have to apply. Torque is in fact the product of two components: *force* and *distance*. You can increase torque by either increasing the applied force, or increasing the distance from the center of rotation. The units of measurement for torque are thus a unit for the force, and a unit for the distance. The International System of Units (SI) defines the newton-meter (Nm) and the newton-centimeter (Ncm).

If you have some familiarity with the properties of levers, you will recognize the similarities. In a lever, the resulting force depends on the distance between the application point and the fulcrum: the longer the distance, the higher the force. You can think of gears as levers whose fulcrum is their axle and whose application points are their teeth. Thus, applying the same force to a larger gear (that is to a longer lever) results in an increase in torque.

When should you gear up or down? Experience will tell you. Generally speaking, you will gear down many more times then you will gear up, because you'll be working with electric motors that have a relatively high velocity yet a fairly low torque. Most of the time, you reduce speed to get more torque and make your vehicles climb steep slopes, or to have your robotic arms lift some load. Other times you don't need the additional torque; you simply want to reduce speed to get more accurate positioning.

One last thing before you move on to the next topic. We said that there is no free lunch when it comes to mechanics. This is true for this conversion service as well: We have to pay something to get the conversion done. The price is paid in *friction*—something you should try and keep as low as possible—but it's unavoid-able. Friction will always eat up some of your torque in the conversion process.

Riding That Train: The Geartrain

The largest LEGO gear is the 40t, while the smallest is the 8t (used in the previous discussion). Thus, the highest ratio we can obtain is 8:40, or 1:5 (Figure 2.2).

Figure 2.2 A 1:5 Gear Ratio

What if you need an even higher ratio? In such cases, you should use a *multi-stage reduction* (or multiplication) system, usually called a *geartrain*. Look at Figure 2.3. In this system, the result of a first 1:3 reduction stage is transferred to a second 1:3 reduction stage. So, the resulting velocity is one third of one third, which is one ninth, while the resulting torque is three times three, or nine. Therefore, the ratio is 1:9.

Figure 2.3 A Geartrain with a Resulting Ratio of 1:9

Geartrains give you incredible power, because you can trade as much velocity as you want for the same amount of torque. Two 1:5 stages result in a ratio of 1:25, while three of them result in 1:125 system! All this strength must be used with care, however, because your LEGO parts may get damaged if for any reason your robot is unable to convert it into some kind of work. In other words, if something gets jammed, the strength of a LEGO motor multiplied by 125 is enough to deform your beams, wring your axles, or break the teeth of your gears. We'll return to this topic later.

Designing & Planning...

Choosing the Proper Gearing Ratio

We suggest you perform some experiments to help you make the right decision in choosing a gearing ratio. Don't wait to finish your robot to discover that some geared mechanics doesn't work as expected! Start building a very rough prototype of your robot, or just of a particular subsystem, and experiment with different gear ratios until you're satisfied with the result. This prototype doesn't need to be very solid or refined, and doesn't even need to resemble the finished system you have in mind. It is important, however, that it accurately simulates the kind of work you're expecting from your robot, and the actual loads it will have to manage. For example, if your goal is to build a robot capable of climbing a slope with a 50 percent grade, put on your prototype all the weight you imagine your final model is going to carry: additional motors for other tasks, the RCX itself, extra parts, and so on. Don't test it without load, as you might discover it doesn't work.

NOTE

Remember that in adding multiple reduction stages, each additional stage introduces further *friction*, the bad guy that makes your world less than ideal. For this reason, if aiming for maximum efficiency, you should try and reach your final ratio with as few stages as possible.

Worming Your Way: The Worm Gear

In your MINDSTORMS box you've probably found another strange gear, a black one that resembles a sort of cylinder with a spiral wound around it. Is this thing really a gear? Yes, it is, but it is so peculiar we have to give it special mention.

In Figure 2.4, you can see a worm gear engaged with the more familiar 24t. In just building this simple assembly, you will discover many properties. Try and turn the axles by hand. Notice that while you can easily turn the axle connected to the worm gear, you can't turn the one attached to the 24t. We have discovered the first important property: The worm gear leads to an *asymmetrical system*; that is, you can use it to turn other gears, but it can't be turned *by* other gears. The reason for this asymmetry is, once again, friction. Is this a bad thing? Not necessarily. It can be used for other purposes.

Figure 2.4 A Worm Gear Engaged with a 24t

Another fact you have likely observed is that the two axles are perpendicular to each other. This change of orientation is unavoidable when using worm gears.

Turning to gear ratios, you're now an expert at doing the math, but you're probably wondering how to determine how many teeth this worm gear has! To figure this out, instead of discussing the theory behind it, we proceed with our experiment. Taking the assembly used in Figure 2.4, we turn the worm gear axle slowly by exactly one turn, at the same time watching the 24t gear. For every turn you make, the 24t rotates by exactly one tooth. This is the answer you were looking for: the worm gear is a 1t gear! So, in this assembly, we get a 1:24 ratio with a single stage. In fact, we could go up to 1:40 using a 40t instead of a 24t.

The asymmetry we talked about before makes the worm gear applicable only in reducing speed and increasing torque, because, as we explained, the friction of this particular device is too high to get it rotated by another gear. The same high friction also makes this solution very inefficient, as a lot of torque gets wasted in the process.

As we mentioned earlier, this outcome is not always a bad thing. There are common situations where this asymmetry is exactly what we want. For example, when designing a robotic arm to lift a small load. Suppose we use a 1:25 ratio made with standard gears: what happens when we stop the motor with the arm loaded? The symmetry of the system transforms the weight of the load (potential energy) into torque, the torque into velocity, and the motor spins back making the arm go down. In this case, and in many others, the worm gear is the proper solution, its friction making it impossible for the arm to turn the motor back.

We can summarize all this by saying that in situations where you desire precise and stable positioning under load, the worm gear is the right choice. And it's also the right choice when you need a high reduction ratio in a small space, since allows very compact assembly solutions.

Limiting Strength with the Clutch Gear

Another special device you should get familiar with is the thick 24t white gear, which has strange markings on its face (Figure 2.5). Its name is *clutch gear*, and in the next part of this section we'll discover just what it does.

Figure 2.5 The Clutch Gear

Our experiment this time requires very little work, just put the end of an axle inside the clutch gear and the other end into a standard 24t to use as a knob. Keep the latter in place with one hand and slowly turn the clutch gear with the other hand. It offers some resistance, but it turns. This is its purpose in life: to offer some resistance, then give in!

This clutch gear is an invaluable help to limit the strength you can get from a geared system, and this helps to preserve your motors, your parts, and to resolve some difficult situations. The mysterious "2.5·5 Ncm" writing stamped on it (as explained earlier, Ncm is a newton-centimeter, the unit of measurement for torque) indicates that this gear can transmit a maximum torque of about 2.5 to 5 Ncm. When exceeding this limit its internal clutch mechanism starts to slip.

What's this feature useful for? You have seen before that through some reduction stages you can multiply your torque by high factors, thus getting a system strong enough to actually damage itself if something goes wrong. This clutch gear helps you avoid this, limiting the final strength to a reasonable value.

There are other cases in which you don't gear down very much and the torque is not enough to ruin your LEGO parts, but if the mechanics jam, the motor stalls—this is a very bad thing, because your motor draws a lot of current and risks permanent damage. The clutch gear prevents this damage, automatically disengaging the motor when the torque becomes too high.

In some situations, the clutch gear can even reduce the number of sensors needed in your robot. Suppose you build a motorized mechanism with a bounded range of action, meaning that you simply want your subsystem (arms, levers, actuators—anything) to be in one of two possible states: open or closed, right or left, engaged or disengaged, with no intermediate position. You need to turn on the motor for a short time to switch over the mechanism from one state to the other, but unfortunately it's not easy to calculate the precise time a motor needs to be on to perform a specific action (even worse, when the load changes, the required time changes, too). If the time is too short, the system will result in an intermediate state, and if it's too long, you might do damage to your motor. You can use a sensor to detect when the desired state has been reached; however, if you put a clutch gear somewhere in the geartrain, you can now run the motor for the approximate time needed to reach the limit in the worst load situation, because the clutch gear slips and prevents any harm to your robot and to your motor if the latter stays on for a time longer than required.

There's one last topic about the clutch gear we have to discuss: where to put it in our geartrain. You know that it is a 24t and can transmit a maximum torque of 5 Ncm, so you can apply here the same gear math you have learned so far. If you place it before a 40t gear, the ratio will be 24:40, which is about 1:1.67. The maximum torque driven to the axle of the 40t will be 1.67 multiplied by 5 Ncm, resulting in 8.35 Ncm. In a more complex geartrain like that in Figure 2.6, the ratio is 3:5 then 1:3, coming to a final 1:5; thus the maximum resulting torque is 25 Ncm. A system with an output torque of 25 Ncm will be able to produce a force five times stronger than one of 5 Ncm. In other words, it will be able to lift a weight five times heavier.

Figure 2.6 Placing the Clutch Gear in a Geartrain

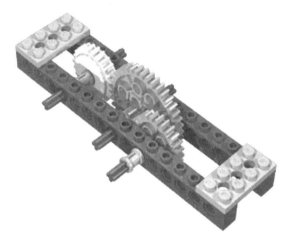

From these examples, you can deduce that the maximum torque produced by a system that incorporates a clutch gear results from the maximum torque of the clutch gear multiplied by the ratio of the following stages. When gearing down, the more output torque you want, the closer you have to place your clutch gear to the source of power (the motor) in your geartrain. On the contrary, when you are reducing velocity, not to get torque but to get more accuracy in positioning, and you really want a soft touch, place the clutch gear as the very last component in your geartrain. This will minimize the final supplied torque.

This might sound a bit complex, but we again suggest you learn by doing, rather than by simply reading. Prototyping is a very good practice. Set up some very simple assemblies to experiment with the clutch gear in different positions, and discover what happens in each case.

Placing and Fitting Gears

The LEGO gear set includes many different types of gear wheels. Up to now, we played with the straight 8t, 24t, and 40t, but the time has come to explore other kinds of gears, and to discuss their use according to size and shape.

The 8t, 24t, and 40t have a radius of 0.5 studs, 1.5 studs, and 2.5 studs, respectively (measured from center to half the tooth length). The distance between the gears' axles when placing them is the sum of their radii, so it's easy to see that those three gears make very good combinations at distances corresponding to whole numbers. 8t to 24t is 2 studs, 8t to 40t is 3 studs, and 24t to 40t equates to four studs. The pairs that match at an even distance are very easy to connect one above the other in our standard grid, because we know it goes by increments of two studs for every layer (Figure 2.7).

Figure 2.7 Vertical Matching of Gears

Another very common straight gear is the 16t gear (Figure 2.8). Its radius is 1, and it combines well with a copy of itself at a distance of two. Getting it to cooperate with other members of its family, however, is a bit more tricky, because whenever matched with any of the other gears it leads to a distance of some studs *and a half*, and here is where the special beams we discussed in the previous chapter (1 x 1, 1 hole, and 1 x 2, 2 holes) may help you (Figure 2.9).

Figure 2.8 The 16t Gear

Figure 2.9 How to Match the 16t Gear to a 24t Gear

Bricks & Chips…

Idler Gears

Figure 2.7 offers us the opportunity to talk about *idler gears*. What's the ratio of the geartrain in the figure? Starting from the 8t, the first stage performs an 8:24 reduction, while the second is a 24:40. Multiplying the two fractions, you get 8:40, or 1:5, the same result you'd get meshing the 8t directly to the 40t. The intermediate 24t is an idler gear, which doesn't affect the gear ratio. Idler gears are quite common in machines, usually to help connect distant axles. Are idler gears totally lacking in effects on the system? No, they have one very important effect: They change the direction of the output!

As we've already said, you're not restricted to using the standard grid. You can try out different solutions that don't require any special parts, like the one showed in Figure 2.10.

Figure 2.10 A Diagonal Matching

When using a pair of 16t gears, the resulting ratio is 1:1. You don't get any effect on the angular velocity or torque (except in converting a fraction of them into friction), but indeed there are reasons to use them as a pair—for instance, when you want to transfer motion from one axle to another with no other effects. This is, in fact, another task that gears are commonly useful for. There's even a class of gears specifically designed to transfer motion from one axle to another axle perpendicular to it, called *bevel gears*.

Designing & Planning...

Backlash

Diagonal matching is often less precise than horizontal and vertical types, because it results in a slightly larger distance between gear teeth. This extra distance increases the *backlash*, the amount of oscillation a gear can endure without affecting its meshing gear. Backlash is amplified when gearing up, and reduced when gearing down. It generally has a bad effect on a system, reducing the precision with which you can control the output axle, and for this reason, it should be kept to a minimum.

The most common member of this class is the 12t bevel gear, which can be used *only* for this task (Figure 2.11), meaning it does not combine at all with any other LEGO gear we have examined so far. Nevertheless, it performs a very useful function, allowing you to transmit the motion toward a new direction, while using a minimum of space. There's also a new 20t bevel conical gear with the same design of the common 12t (Figure 2.12). Both of these bevel gears are half a stud in thickness, while the other gears are 1 stud.

Figure 2.11 Bevel Gears on Perpendicular Axles

Figure 2.12 The 20t Bevel Gear

The 24t gear also exists in the form of a *crown gear*, a special gear with front teeth that can be used like an ordinary 24t, which can combine with another straight gear to transmit motion in an orthogonal direction (that is, composed of right angles), possibly achieving at the same time a ratio different from 1:1 (Figure 2.13).

To conclude our discussion of gears, we'll briefly introduce some recent types not included in the MINDSTORMS kit, but that you might find inside other LEGO sets. The two *double bevel* ones in Figure 2.14 are a 12t and a 20t, respectively 0.75 and 1.25 studs in radius. If you create a pair that includes one per kind of the two, they are an easy match at a distance of 2 studs.

Figure 2.13 The Crown Gear on Perpendicular Axles

Figure 2.14 Double Bevel Gears

Things get a bit more complicated when you want to couple two of the same kind, as the resulting distance is 1.5 or 2.5. And even more complicated when combined with other gears, causing resulting distances that include a quarter or three quarters of a stud. These gears are designed to work well in perpendicular setups as well (Figure 2.15).

Figure 2.15 Double Bevel Gear on Perpendicular Axles

Using Pulleys, Belts, and Chains

The MINDSTORMS kit includes some *pulleys* and *belts*, two classes of components designed to work together and perform functions similar to that of gears—

similar, that is, but not identical. They have indeed some peculiarities which we shall explore in the following paragraphs.

Chains, on the other hand, are not part of the basic MINDSTORMS kit. You will need to buy them separately. Though not essential, they allow you to create mechanical connections that share some properties with both geartrains and pulley–belt systems.

Pulleys and Belts

Pulleys are like wheels with a groove (called a *race*) along their diameter. The LEGO TECHNIC system currently includes four kinds of pulleys, shown in Figure 2.16.

Figure 2.16 Pulleys

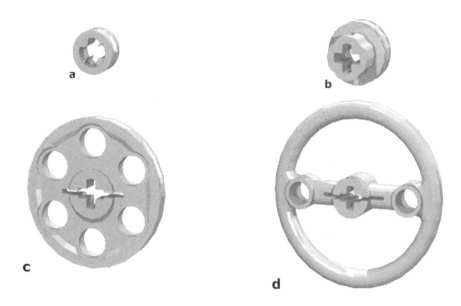

The smallest one (a) is actually the half-size bush, normally used to hold axles in place to prevent them from sliding back and forth. Since it does have a race, it can be properly termed a pulley. Its diameter is one LEGO unit, with a thickness of half a unit.

The small pulley (b) is 1 unit in thickness and about 1.5 units in width. It is asymmetrical, however, since the race is not in the exact center. One side of the axle hole fits a rubber ring that's designed to attach this pulley to the micro-motor. The medium pulley (c) is again half a unit thick and 3 units in diameter. Finally, the large pulley (d) is 1 unit thick and about 4.5 units in diameter.

LEGO belts are rings of rubbery material that look similar to rubber bands. They come in three versions in the MINDSTORMS kit, with different colors corresponding to different lengths: white, blue, and yellow (in other sets, you can find a fourth size in red). Don't confuse them with the actual rubber bands, the black ones you found in the kit: Rubber bands have much greater elasticity, and for this reason are much less suitable to the transfer of motion between two pulleys. This is, in fact, the purpose of belts: to connect a pair of pulleys. LEGO belts are designed to perfectly match the race of LEGO pulleys.

Let's examine a system made of a pair of pulleys connected through a belt (Figure 2.17). The belt transfers motion from one pulley to the other, making them similar to a pair of gears. How do you compute the ratio of the system? You don't have any teeth to count... The rule with pulleys is that the reduction ratio is determined by finding the ratio between their diameters (this rules applies to gears too, but the fact that their circumference is covered with evenly spaced teeth provides a convenient way to avoid measurement). You actually should consider the diameter of the pulley *inside* its race, because the sides of the race are designed specifically to prevent the belt from slipping from the pulley and don't count as part of the diameter the belt acts over.

Figure 2.17 Pulleys Connected with a Belt

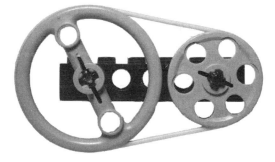

You must also consider that pulleys are not very suitable to transmitting high torque, because the belts tend to slip. The amount of slippage is not easy to estimate, as it depends on many factors, including the torque and speed, the tension of the belt, the friction between the belt and the pulley, and the elasticity of the belt.

For those reasons, we preferred an experimental approach and measured some actual ratios among the different combination of pulleys under controlled conditions. You can find our results in Table 2.1.

Table 2.1 Ratios Among Pulleys

	Half Bush	Small Pulley	Medium Pulley	Large Pulley
Half bush	1:1	1:2	1:4	1:6
Small pulley	2:1	1:1	1:2.5	1:4.1
Medium pulley	4:1	2.5:1	1:1	1:1.8
Large pulley	6:1	4.1:1	1.8:1	1:1

Designing & Planning...

Finding the Ratio between Two Pulleys

How did we find out the actual ratio between two pulleys? By simply connecting them with a belt and comparing the number of rotations when one of the two gets turned and drags the other. But turning pulleys by hand would have been quite a boring and time-consuming task, and could cause some counts to be missed. What better device for this job than our RCX, equipped with a motor and two rotation sensors? So, we built this very simple machine: a motor connected to a pulley, whose axle is attached to the first rotation sensor, and a second pulley, placed at a very short distance, with its axle attached to the second rotation sensor. We used some care to minimize the friction and maintain the same tension in the belt for all the pairs of pulleys.

When running the motor, the RCX counted the rotations for us. We stopped the motor after a few seconds, read the rotation sensor counts, and divided the two to get the ratio you see in Table 2.1.

These values may change significantly in a real-world application, when the system is under load. Because of this, it's best to think of the figures as simply an indication of a possible ratio for systems where very low torque is applied. Generally speaking, you should use pulleys in your first stages of a reduction system, where the velocity is high and the torque still low. You could even view the slippage problem as a positive feature in many cases, acting as a torque-limiting mechanism like the one we discussed in the clutch gear, with the same benefits and applications.

Another advantage of pulleys over gear wheels is that their distance is not as critical. Indeed, they help a great deal when you need to transfer motion to a distant axle (Figure 2.18). And at high speeds they are much less noisy than gears—a facet that occasionally comes in handy.

Figure 2.18 Pulleys Allow Transmission across Long Distances

Chains

LEGO *chains* come in two flavors: *chain links* and *tread links* (as shown in Figure 2.19, top and bottom, respectively). The two share the same hooking system and are freely mixable to create a chain of the required length.

Chains are used to connect gear wheels as the same way belts connect with pulleys. They share similar properties as well. Both systems couple parallel axles without reversing the rotation direction, and both give you the chance to connect distant axles. The big difference between the two is that chain links don't allow any slippage, so they transfer *all* the torque. (The maximum torque a chain can transfer depends on the resistance of its individual links, which in the case of LEGO chains is not very high.) On the other hand, they introduce further friction into the system, and for this reason are much less efficient then direct gear matches. You will find chains useful when you have to transfer motion to a distant axle in low velocity situations. The ratio of two gears connected by a chain is the same as their corresponding direct connection. For example, a 16t connected to a 40t results in a 2:5 ratio.

Figure 2.19 Chain Links

Making a Difference: The Differential

There's a very special device we want to introduce you to at this time: the *differential gear*. You probably know that there's at least one differential gear in every car. What you might not know is why the differential gear is so important.

Let's do an experiment together. Take the two largest wheels that you find in the MINDSTORMS kit and connect their hubs with the longest axle (Figure 2.20). Now put the wheels on your table and push them gently: They run smoothly and advance some feet, going straight. *Very straight*. Keep the axle in the middle with your fingers and try to make the wheels change direction while pushing them. It's not so easy, is it?

Figure 2.20 Two Connected Wheels Go Straight

The reason is that when two parallel wheels turn, their paths must have different lengths, the outer one having a longer distance to cover (Figure 2.21). In our example, in which the wheels are rigidly connected, at any turn they cover the same distance, so there's no way to make them turn unless you let one slip a bit.

Figure 2.21 During Turns the Wheels Cover Different Distances

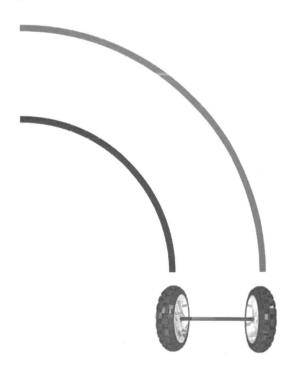

The next phase of our experiment requires that you now build the assembly shown in Figure 2.22. You see a differential gear with its three 12t bevel gears, two 6–stud axles, and two beams and plates designed to provide you with a way to handle this small system. Placing the wheels again on your table, you will notice that while pushing them, you can now easily turn smoothly in any direction. Please observe carefully the *body* of the differential gear and the central bevel gear: when the wheels go straight, the body itself rotates while the bevel gear is stationary. On the other hand, if you turn your system in place, the body stays put and the bevel gear rotates. In any other intermediate case, both of them rotate at some speed, adapting the system to the situation. Differentials offer a way to put power to the wheels without the restriction of a single fixed drive axle.

Figure 2.22 Connecting Wheels with the Differential Gear

To use this configuration in a vehicle, you simply have to apply power to the body of the differential gear, which incorporates a 24t on one side and a 16t on the other.

The differential gear has many other important applications. You can think of it as a mechanical adding/subtracting device. Again place the assembly from Figure 2.22 on your table. Rotate one wheel while keeping the other from turning; the body of the differential gear rotates half the angular velocity of the rotating wheel. You already discovered that when turning our system in place, the differential does not rotate at all, and then when both wheels rotate together, the differential rotates at the same speed as well. From this behavior, we can infer a simple formula:

(Iav1 + Iav2) / 2 = Oav

where *Oav* is the *output angular velocity* (the body of the differential gear), and *Iav1* and *Iav2* are the *input angular velocities* (the two wheels). When applying this equation, you must remember to use *signed* numbers for the input, meaning that if one of the input axles rotates in the opposite direction of the other, you must input its velocity as a negative number. For example, if the right axle rotates at 100 rpm (revolutions per minute) and the left one at 50 rpm, the angular velocity of the body of the differential results in this:

(100 rpm + 50 rpm) / 2 − 75 rpm

There are situations where you deliberately reverse the direction of one input, using idler gears, to make the differential sensitive to a difference in the speed of the wheels, rather than to their sum. Reversing the input means that you must make one of the inputs negative. See what happens to the differential when both wheels run at the same speed, let's say 100 rpm:

(100 rpm − 100 rpm) / 2 = 0 rpm

It doesn't move! As soon as a difference in speed appears, the differential starts rotating with an angular velocity equal to half this difference:

(100 rpm − 98 rpm) / 2 = 1 rpm

This is a useful trick when you want to be sure your wheels run at the same speed and cover the same distance: Monitor the body of the differential and slow the left or right wheel appropriately to keep it stationary. See Chapter 8 for a practical application of this trick.

Summary

Few pieces of machinery can exist without gears, including robots, and you ought to know how to get the most benefit from them. In this chapter, you were introduced to some very important concepts: gear ratios, angular velocity, force, torque, and friction. Torque is what makes your robot able to perform tasks involving force or weight, like lifting weights, grabbing objects, or climbing slopes. You discovered that you can trade off some velocity for some torque, and that this happens along rules similar to those that apply to levers: the larger the distance from the fulcrum, the greater the resulting force.

The output torque of a system, when not properly directed to the exertion of work, or when something goes wrong in the mechanism itself, can cause damage to your LEGO parts. You learned that the clutch gear is a precious tool to limit and control the maximum torque so as to prevent any possible harm.

Gears are not the only way to transfer power; we showed that pulley-belt systems, as well as chains, may serve the same purpose and help you in connecting distant systems. Belts provide an intrinsic torque-limiting function and do well in high-speed low-torque situations. Chains, on the other hand, don't limit torque but do increase friction, so they are more suitable for transferring power at slow speeds.

Last but not least, you explored the surprising properties of the differential gear, an amazing device that can connect two wheels so they rotate when its body rotates, still allowing them to turn independently. The differential gear has some other applications, too, since it works like an adder-subtracter that can return the algebraic sum of its inputs.

If these topics were new to you, we strongly suggest you experiment with them before designing your first robot from scratch. Take a bunch of gears and axles and play with them until you feel at ease with the main connection schemes and their properties. This will offer you the opportunity to apply some of the concepts you learned from Chapter 1 about bracing layers with vertical beams to make them more solid (when you increase torque, many designs fall apart unless properly reinforced). You won't regret the time spent learning and building on this knowledge. It will pay off, with interest, when you later face more complex projects.

Chapter 3

Controlling Motors

Solutions in this chapter:

- Pacing, Trotting, and Galloping
- Mounting Motors
- Wiring Motors
- Controlling Power
- Coupling Motors

Introduction

Motors will be your primary source of power. Your robots will use them to move around, lift loads, operate arms, grab objects, pump air, and perform any other task that requires power. There are different kinds of electric motors, all of them sharing the property of converting *electrical energy* into *mechanical energy*. In this chapter, we will survey different kinds of LEGO motors and will discuss how to use, mount, connect, and combine them.

Before entering the world of motors, we would like to introduce you to some basic concepts about electricity. There's a very important distinction you should be aware of concerning electrical current: There are two types, *alternating current* (AC) and *direct current* (DC). Alternating current is the type of electricity that comes out of the wall outlets in your house, while batteries are the most typical source of direct current. All the electric LEGO devices, including motors, work with DC only.

To understand what DC is, imagine a stream of water going down a hill. Electricity flowing through a wire is not very different: When you connect a battery to a device like a lamp or a motor, you enable a circuit through which electricity flows more or less like water in a stream. You know that batteries have positive (+) and negative (−) signs stamped on them: they indicate the direction of the flow, which goes from minus to plus, as if the minus pole were the top of the hill. You can place a water mill along the stream to convert the energy of water into mechanical energy; similarly, an electric motor converts an electric flow into motion. What would happen to the water mill if you could reverse the direction of the stream? It would change its direction of rotation. The same happens to DC motors. Every motor has two connectors, one to attach to the negative pole and the other to connect to the positive end of a DC source. You can imagine the current flowing from the negative pole of the battery into the motor, making it move and then coming out again to return to the positive pole of the battery. If you reverse the *polarity*, that is, if you swap the wires between the motor and the battery, you will change the direction of the stream and thus the direction of the motor.

Continuing with our hydraulics metaphor, how would you describe the *quantity* of water that's flowing in a stream? It depends on two factors: the speed of the water, and the width of the stream. Both of them have an influence on the kind of work your mill can perform. In the realm of electricity, the speed of the stream is called *voltage*, and its width (its intensity) is called *current*. They are respectively expressed in Volts (V) and Ampere (A), or sometimes in their submultiples, *millivolt* (mV) and *milliampere* (mA). The amount of work that an electrical flow can perform, for example through a motor, depends on both these quantities. To be more precise, it depends on their product, called *power*, and is measured in Watts (W).

Every motor is designed to run at a specific voltage, but they are very tolerant when it comes to decreases in the supplied voltage. They simply turn more slowly. However, if you *increase* the voltage above the specific limit for a motor, you stand a good chance of burning it out.

Current has a different behavior. It's the motor that "decides" how much current to draw according to the work it's doing: the higher the load, the greater the current. The situation you should avoid at all costs when working with your RCX is to have the motor *stall* (it is connected to the power source but something prevents it from turning). What happens in this case is that the motor tries to win out against the resistance, drawing in more current so it can convert it into power, but as it doesn't succeed in the task, all that current becomes *thermal energy* instead of mechanical energy—in other words, heat. This is the most dangerous condition for an electric motor. And here is where the clutch gear described in Chapter 2 comes into play, limiting the maximum torque and thus preventing stall situations. You will discover later in the chapter that the RCX also has an active role in protecting your motors from dangerous draws of current.

Pacing, Trotting, and Galloping

Every motor contains one or more coils and permanent magnets that convert electrical energy into mechanical energy, but you don't really need to know this level of detail. What you, as a robot builder, must remember is that every motor has a connector through which you can supply it energy, and an output *shaft* which draws the power. The current LEGO TECHNIC line includes three types of 9V DC motors (as shown in Figure 3.1): the ungeared motor (a), the geared motor (b), and the micromotor (c). There are other special motors as well: the train motor, the geared motor with battery pack, and the Micro Scout unit. These are less common, less useful, and less versatile to robotics than the first three, so we won't be examining them here. Table 3.1 summarizes the properties of these three motors.

Table 3.1 Properties of the LEGO TECHNIC Motors

Properties	Ungeared Motor	Geared Motor	Micromotor
Maximum voltage	9V DC	9V DC	9V DC
Minimum current (no load)	100 mA	10 mA	5 mA
Maximum current (stall)	450 mA	250 mA	90 mA
Maximum speed (no load)	4000 rpm	350 rpm	30 rpm
Speed under typical load	2500 rpm	200 rpm	25 rpm

Figure 3.1 The LEGO TECHNIC Motors

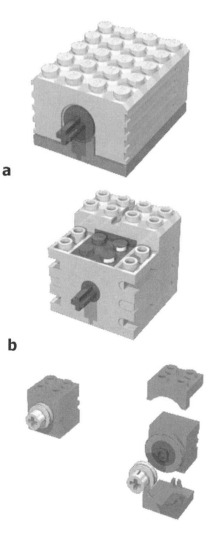

a

b

c

The ungeared motor (a) has been the standard LEGO TECHNIC motor for a long time. Its axle is simply an extension of the inner electric motor shaft, and for this reason we called it *ungeared*. Electric motors usually rotate at very high speeds, and this one is no exception, turning at more then 4000 rpm (revolutions per minute). This makes this motor a bit tricky to use, because it requires very high reduction ratios for almost any practical application, leading to very cumbersome and complex geartrains. Add the fact that it draws an amazing amount of current, and you get a pretty good picture of how difficult it can be.

This motor is still easy to find in the shops of many countries as an expansion pack (8720), but we strongly advise you against buying one for the reasons mentioned in the previous paragraph. In this book, you won't find any example that includes the ungeared motor. Nevertheless, if you already have one, you can safely use it; it won't damage your RCX or be damaged itself. The only risk you're taking is that, under heavy loads or stall situations, it drains your batteries very quickly.

The geared motor (b) is what we will generically refer to as a *motor*, the one we will use extensively in the following chapters. It features an internal multistage reduction geartrain and turns at about 350 rpm with no load (typically 200/250rpm with medium load). It's much more efficient than the older kind, and has low current consumption. It also uses more compact geartrains. If you have the MINDSTORMS kit, you already have two of these.

Bricks and Chips…

How to Release a Jammed Micromotor

A micromotor jams so easily, you should know what to do when it occurs. The following list should help:

1. Switch off the motor as soon as you can. Detach the cord or switch the power off; it's important not to leave a stalled motor under power for a long time because that could permanently damage it.

2. Decouple the motor from any connection (gearings, pulleys, and so on). Leave the small pulley attached to the motor shaft.

3. Holding the motor with your fingers, turn the pulley gently but firmly in the same direction the motor was turning when

Continued

> it jammed. At the same time, push the pulley against the motor until you hear a click. Your motor should be okay now. If you don't know what direction the motor was rotating when jammed, try both directions.
>
> This procedure usually works. If it doesn't, try to power on the motor in both directions with very brief current pulses, at the same time doing what's described in Step 3.

The micromotor (c) is a geared motor as well. It's geared down so much that its output shaft turns at approximately 30 rpm. Nevertheless, its torque is incredibly low, well below 1 Ncm. It is also surprisingly noisy, and very easy to jam. At this point, you might wonder why you should ever consider this motor, but the answer lies in its name: because it's micro. There are situations where the size of the motor is more critical than the amount of torque and speed needed. To be used, it requires some special mounting brackets, and a small pulley to connect to its shaft (Figure 3.1c).

Mounting Motors

The LEGO motor is 4 studs wide and 4 studs long, and has an irregularly shaped top with a height of 2.33 bricks in the low part and 3 bricks in the high one. The base of the motor is irregular too, because there's a convex area of 2 x 2 that makes a direct mount of the motor on a regular surface impossible. For these reasons mounting LEGO motors requires some experience. In the following paragraphs, we'll show a few common solutions, but many others will work as well.

Despite its irregular shape, the motor fits well enough in the standard grid. In Figure 3.2, you see that its lower part can be locked between two beams at a distance of four holes. It is very important that you build your motors inside a solid structure, otherwise they will become loose when you apply a load. You can also see in the figure that the shaft of the motor is two holes from the bottom beam, which is perfect for some of the gearing combinations discussed in the previous chapter: 8t to 24t, or 16t to 16t, to name a few.

In our second example (displayed in Figure 3.3), we show another very solid assembly. This time we extended the output axle of the motor in order to mount a worm gear on it so it can mesh with a 24t. While the previous case was suitable to drive wheels from the 24t axle, this would fit a slow speed/high torque application.

Figure 3.2 Locking a Motor between Beams

Figure 3.3 A Motor Connected to a Worm Gear

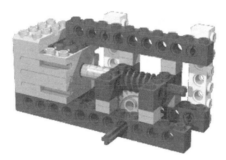

NOTE

The pictures here are mainly meant to highlight relations and distances. So, in order to let you see inside, we didn't lock the motor on both sides. In actual applications, you will complete the assembly and adjust the beams to the proper length for your needs.

Your MINDSTORMS kit contains eight 1 x 2 plates with rails which are specifically designed as brackets for the motors (Figure 3.4). They permit compact and solid attachments like the one in Figure 3.5. But even more importantly, they give you the ability to remove motors without dismantling your robot. In the example in Figure 3.5, if you remove the two 2 x 6 plates behind the motor, you can easily slip it off without altering the rest of the assembly. This is a very desirable property, allowing you to recycle your motors in other projects without being forced to take your robot apart. You will likely end up having more LEGO parts than those contained in your MINDSTORMS kit, so it's possible you'll have more than a single assembled robot at one time. Motors are among the most expensive LEGO components. Reusing them in different projects will help keep the cost of your hobby at a reasonable level!

Figure 3.4 1 x 2 Plates with Rails Provide a Convenient Mounting Solution

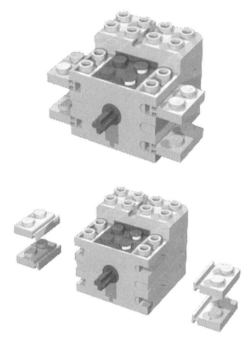

Figure 3.5 An Easily Removable Motor

NOTE

We suggest that, when mounting motors, you keep the wire free to be removed. Don't block it together with the motor, unless you're sure your design won't change and you won't need a wire of different length.

Figure 3.6 illustrates our last example. You can see how two pulleys and a belt may solve the problem of transferring power to a distant axle through a narrow space. In this particular example, the motor does not need to be locked with a vertical beam because the torque on its shaft won't ever reach high values (belt slippage prevents this from happening). At the same time, the belt works like a rubber band, too, keeping the motor from coming off its foundation.

Figure 3.6 Belts Don't Require Very Solid Mountings

Wiring Motors

The LEGO wiring system is so easy to use you won't require any training. The cables end with 2 x 2 x 2/3 connectors that attach as easily as standard bricks and don't need any special knowledge to be used.

As we already explained, LEGO motors are DC motors, therefore they are sensitive to the *polarity* you connect them with, meaning it determines whether the motor turns clockwise or counterclockwise. Usually, you don't have to worry about this, since you can control this property from your program. However, the design of the LEGO connectors is very clever and not only prevents you from involuntarily short-circuiting the motor or the battery, but they also allow you to reverse the polarity by simply turning them 180 degrees.

How can you test your motors without adjusting programming? There are many different ways, as in the following:

- **RCX console** Press the **View** button until you select the port your motor is wired to. When the cursor (a small arrow) points to the proper port, don't release the button. Keeping the **View** button pressed, you can press **Prgm** or **Run** to power the motor in the desired direction.

- **Software** Browsing the Internet you can find and download many good freeware programs that allow full direct control of your RCX via your PC. They make running a motor as easy as a click of the mouse (see Appendix A for links and resources).

- **External battery box** Some LEGO TECHNIC sets include a battery box (Figure 3.7). If you want an extra motor and buy an 8735 TECHNIC Motor set, you'll get one. With this box you can test your motor with no need of the RCX.

Figure 3.7 The LEGO Battery Box

- **Remote control** This useful tool is not included in the MIND-STORMS kit, you have to buy it separately (Figure 3.8). It's currently sold inside the Ultimate Accessory Set that also contains additional parts. If you can afford it, it's a good buy. You can control all three output ports at the same time, which is very useful when testing your robot during the building phase.

- **Other sources** All the components of the LEGO 9V electric system are compatible with each other. If you have a LEGO train speed regulator, or a Control Center unit, you can safely use them to run your motors. Don't use non-LEGO electricity sources. They might harm your motors.

Figure 3.8 The LEGO MINDSTORMS Kit Remote Control

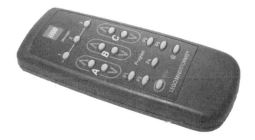

In some cases, you want to control more than a single motor from the same RCX output port. Is this safe for your RCX and your motors? Yes, and with no risk of damaging either item. The only thing to point out is that the RCX has a current-limiting device behind each port that prevents your motor from drawing too much current to avoid any possible damage during stall situations. When you connect two or more motors to the same port, they must share the maximum available current, thus limiting the work they can perform. Nevertheless, there are situations where splitting the load on two or more motors is the preferable option.

There is another possible approach that bypasses the current-limiting circuit: *indirect control*. Instead of supplying the motors from your RCX port, you control a motor that activates a switch that turns on the other motors. This sounds complicated, but it isn't. You just need some extra parts: a polarity switch and a battery box. In Figure 3.9, you see a system devised to drive the LEGO polarity switch with a motor and two pulleys. The belt coupling makes the system less critical about timing. If you accidentally power the controlling motor for longer than what's needed to activate the switch, the belt slips and your motor doesn't stall.

The polarity switch is actually a three-state switch: *forward*, *off*, and *reverse*. At one side, it switches the motors on, in the center it switches them off, while on the other side it switches them on again but with reversed polarity. Our simple assembly can control only two states (don't rely on timing to position the polarity switch precisely in the center!), so you have to choose whether you want an on/off system or a forward/reverse one.

As the battery box does not feature any current limiting device, your motors can draw as much current as they need out of the batteries. Remember that with this wiring the controlled motors are not protected against overloads, thus stall situations might permanently damage them.

Figure 3.9 Indirect Motor Control

Controlling Power

You know that your program can control the power of your motors. In fact, a specific instruction will set the power level in the range 0 to 7 (some alternative firmware, like legOS, provide higher granularity, e.g., 0 to 255). But what happens when you change this number? And why do we care?

There are different ways to control the power of an electric motor. The LEGO train speed regulator controls power through voltage: the higher the voltage, the higher the power. The RCX uses a different approach, called *pulse width modulation* (PWM).

To explain how this works, imagine that you continuously and rapidly switch your motor on and off. The power your motor produces in any given interval depends on how long it's been on in that period. Applying current for a short period of time (a *low duty cycle*) will do less work than applying it for a longer time. If you could switch it on and off hundreds of times a second, you would see the motor turning in an apparently normal way; but under load you would notice a decrease in its speed, due to a decrease in the supplied power (Figure 3.10).

This is exactly what the RCX does. Its internal motor controller can switch the power on and off very quickly (an on/off cycle every 8 milliseconds), at the same time varying the proportion between the on period and the off period. At power level 0, the motor is on for 1/8 of the cycle; at power level 1, for 2/8 of it; and so on until you reach level 7, when the motor is always on (8/8).

Figure 3.10 Pulse Width Modulation Power Levels

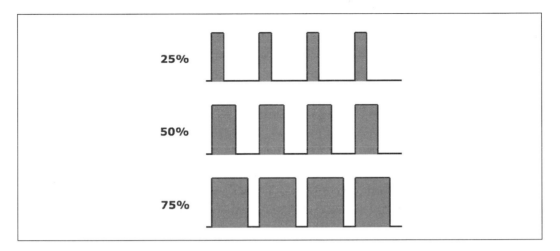

Why do we care about this technical stuff? Because this explains you aren't actually controlling speed, but power. LEGO motors are very efficient, and when the motor has no load or a very small one, lowering the power level won't decrease its speed very much. Under more load, you will see how the power level affects the resulting speed, too.

Braking the Motor

Controlling the power means also being able to brake your motor when necessary. For this purpose, the RCX features a sort of electric brake. Once again, let us explain how it works through an experiment.

You need a motor, a cable (any length), and a 24t gear. Assemble the three as shown in Figure 3.11, paying attention to the way the cable is looped: the ends of the wire go on opposite sides. Now try and turn the 24t with your fingers: it turns smoothly, and continues to spin for a while after you've stopped turning it.

Then remove the cable and reconnect it as shown in Figure 3.12: the ends of the wire go into the same side—this way the motor is short-circuited. We know that a *short circuit* sounds like a *bad* thing, but in this particular case we mean only that the circuit is *closed*. Don't worry, your motor is not at any risk. Now try and turn the 24t again. You see? The motor offers a lot of resistance, and as soon as you stop turning, it stops, too.

What happened? A LEGO motor is not only able to transform electricity into motion, it does the opposite, too: It can be used to *generate* electricity. In our experiment the generated current short-circuits back into the motor, producing the force that resists the motion. This is the simple but effective system the RCX implements to brake the motor: When you set them to *off*, the RCX not only switches the power off, it also short-circuits the port, making the motor brake.

Figure 3.11 In This Setup, the Motor Shaft Turns Smoothly

Figure 3.12 An Electric Brake

There's another condition, called *float* mode, where the RCX simply disconnects the motor without creating any brake effect. In this case, the motor will continue to turn for a few seconds after the power has been removed.

Using Motors as Generators

If you are not convinced that a motor works as a generator, too, perform this simple experiment. Connect one motor to another with a wire. Place a 24t on each shaft. Take one motor in your hands and turn the 24t while looking at the second motor. What happens? The first motor converts the mechanical energy coming from your fingers into electric current, which makes the second motor turn.

Coupling Motors

We previously discussed the case in which you want to wire two motors to the same port. If you do this to get more power for a task, you will very likely need to mechanically *couple* the motors as well, meaning that they will work together to operate the same mechanism, sharing its load. It's like when you have to move something really heavy and call a friend to help you: each member of the party bears only half the total weight. Though this rule works for all electric motors in general, a specific limitation applies when attaching LEGO motors to the RCX: Its current-limiting device won't allow the motors to draw as much current as they want. Consider it a constraint to the maximum power each port can pay out.

In Figure 3.13, you see two motors acting upon the same 40t gear wheel. People often wonder whether connections like these are going to cause any problem to the motors. The answer is simply *no*. Unless you keep your motor stalled for more than a brief moment, they are not easy to damage. In applications like the one in Figure 3.13, you just have to be sure the motors don't oppose each other. With this in mind, we suggest you double-check both the connection and turning directions before actually coupling the motors to the same gear.

It is true that no two motors turn exactly at the same speed, or output the same torque either, but this doesn't cause any conflict. A motor doesn't *know* that there's another motor cooperating on the same task, it simply reacts to the load absorbing more current and trying to keep the speed. This works even if the motors are of different types, even if they are powered at different levels, and even if they are geared with different ratios.

If you're not convinced of this, think of a simple vehicle propelled by a single motor. When the path becomes steeper, the load on the motor increases, causing it to reduce its speed. Essentially, the motor adapts itself to the load. The same happens when two motors work together, they share the load and mutually adapt themselves.

Figure 3.13 Two Mechanically Coupled Motors

Have you ever tried riding a tandem bicycle? Your partner might be much weaker than you, but you would prefer him to pedal rather than simply ride along watching the landscape.

Summary

LEGO electric motors are easy and safe to use, but they require a bit of experience to get the most from them and avoid any possible damage. On this latter topic, the most important thing is to never let them stall for more than a few seconds and to never keep them powered when they've stalled. You already know from Chapter 2 that the clutch gear is a good ally in this venture, and you've now learned that the RCX has further protections that limit the maximum current and thus the risk that your motor will burn out.

You've seen that wiring LEGO motors is very simple: The special connectors prevent short circuits and allow easy control of polarity, which affects the direction in which a motor turns. The different mounting options require a bit of practice, the same as for gears. Don't forget to brace motors with vertical beams the way you were taught in Chapter 1: They produce enough torque to pull themselves apart if not solidly locked!

On the topic of coupling motors, this option is useful when you want to split a load over two or more of them to reduce their individual effort. The only important thing to remember is that you must control them from the same port, so as to avoid any dangerous conflict situation where one motor opposes to the other.

As a general tip, we suggest you make intense use of prototyping—don't wait to finish your robot to discover a motor is in the wrong place or has not been geared properly—test your mechanisms while you are building them.

Chapter 4

Reading Sensors

Solutions in this chapter:

- Touch Sensor
- Light Sensor
- Rotation Sensor
- Temperature Sensor
- Sensor Tips and Tricks
- Other Sensors

Introduction

Motors, through gears and pulleys, provide motion to your robot; they are the muscles that move its legs and arms. The time has come to equip your creature with *sensors*, which will act as its eyes, ears, and fingers.

The MINDSTORMS box contains two types of sensors: the *touch sensor* (two of them) and the *light sensor*. In this chapter, we'll describe their peculiarities, and those of the optional sensors that you can buy separately: the *rotation sensor* and the *temperature sensor*. All these devices have been designed for a specific purpose, but you'll be surprised at their versatility and the wide range of situations they can manage. We will also cover the cases where one type of sensor can *emulate* another, which will help you replace those that aren't available. Using a little trick that takes advantage of the infrared (IR) light on the RCX, you will also discover how to turn your light sensor into a sort of radar.

We invite you to keep your MINDSTORMS set by your side while reading the chapter, so you can play with the real thing and replicate our experiments. For the sake of completeness, we'll describe some parts that come from MIND-STORMS expansion sets or TECHNIC sets. Don't worry if you don't have them now; this won't compromise your chances to build great robots.

Touch Sensor

The *touch* sensor (Figure 4.1) is probably the simplest and most intuitive member of the LEGO sensor family. It works more or less like the push button portion of your doorbell: when you press it, a circuit is completed and electricity flows through it. The RCX is able to detect this flow, and your program can read the state of the touch sensor, **on** or **off**.

Figure 4.1 The Touch Sensor

If you have already played with your RIS, read the Constructopedia, and built some of the models, you're probably familiar with the sensors' most common application, as *bumpers*. Bumpers are a simple way of interacting with the environment; they allow your robot to detect obstacles when it hits them, and to change its behavior accordingly.

A bumper typically is a lightweight mobile structure that actually hits the obstacles and transmits this impact to a touch sensor, closing it. You can invent many types of bumpers, but their structure should reflect both the shape of your robot as well as the shape of the obstacles it will meet in its environment. A very simple bumper, like the one in Figure 4.2, could be perfectly okay for detecting walls, but might not work as expected in a room with complex obstacles, like chairs. In such cases, we suggest you proceed by experimenting. Design a tentative bumper for your robot and move it around your room at the proper height from the floor, checking to see if it's able to detect all the possible collisions. If your bumper has a large structure, don't take it for granted that it will impact the obstacle in its optimal position to press the sensor. Our example in Figure 4.2 is actually a bad bumper, because when contact occurs, it hardly closes the touch sensors at the very end of the traverse axle. It's also a bad bumper because it transmits the entire force of the collision straight to the switch, meaning an extremely solid bracing would be necessary to keep the sensor mounted on the robot.

Figure 4.2 A Simple Bumper

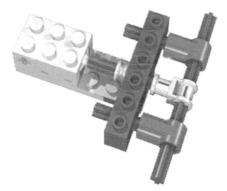

Be empirical, try different possible collisions to see if your bumper works properly in any situation. You can write a very short program that loops forever, producing a beep when the sensor closes, and use it to test your bumper.

When talking of bumpers, people tend to think they should *press* the switch when an obstacle gets hit. But this is not necessarily true. They could also *release* the switch during a collision. Look at Figure 4.3, the rubber bands keep the bumper gently pressed against the sensor; when the front part of the bumper touches something, the switch gets released.

Figure 4.3 A Normally Closed Bumper

Actually, there are some important reasons to prefer this kind of bumper:

- The impact force doesn't transfer to the sensors itself. Sensors are a bit more delicate than standard LEGO bricks and you should avoid shocking them unnecessarily.

- The rubber bands absorbing the force of the impact preserve not only your sensor but the whole body of your robot. This is especially important when your robot is very fast, very heavy, very slow in reacting, or possesses a combination of these factors.

Bumpers are a very important topic, but touch sensors have an incredible range of other applications. You can use them like buttons to be pushed manually when you want to inform your RCX of a particular event. Can you think of a possible case? Actually, there are many. For example, you could push a button to order your RCX to "read the value of the light sensor *now*," and thus calibrate readings (we will discuss this topic later). Or you could use two buttons to give feedback to a learning robot about its behavior, *good* or *bad*. The list could be long.

Another very common task you'll demand from your switch sensors is *position control*. You see an example of this in Figure 4.4. The rotating head of our robot

(Figure 4.4a) mounts a switch sensor that closes when the head looks straight ahead (Figure 4.4b). Your software can rely on timing to rotate the head at some level (right or left), but it can always drive back the head precisely in the center simply waiting for the sensor to close. By the way, the *cam* piece we used in this example is really useful when working with touch sensors, as its three half-spaced crossed holes allow you to set the proper distance to close the sensor.

Figure 4.4 Position Control with a Touch Sensor

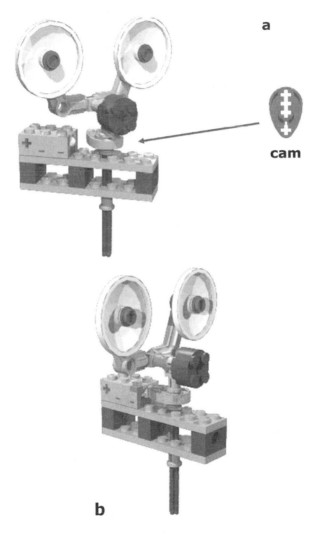

There would be many other possible applications in regards to position control. We'll meet some of them in the third part of this book. What matters here is to invite you to explore many different approaches before actually building your robot. Let's create another example to clarify what we mean. Suppose you're going to build an elevator. You obviously want your elevator to stop at any floor. Your first idea is to put a switch at every level, so when one of them closes you know that the cab has reached that level. Okay, nice approach. There's one small problem; however, you have just two touch sensors, and an elevator with only two floors doesn't seem like such an interesting project to you. You could buy a third sensor, but this simply pushes your problem one floor up, without solving the general case. Meanwhile, the three input ports of your RCX are all engaged. Suddenly, an idea occurs to you: Why not put the sensor on the booth instead of on the structure? With a single sensor on the booth, and pegs that close it at any floor, you can provide your elevator with as many floors as you like. You see, by reversing our original approach you found a much better solution. Are the two systems absolutely equivalent? No, they aren't. In the first, you could determine the absolute position of the booth, while in the second you are able to know only its relative position. That is, you do need a known starting point, so you can deduce the position of the cab counting the floors from there. Either require that the cab must be at a specific level when the program starts, or use a second sensor to detect a specific floor. For example, place a sensor at the ground level, so the very first thing your program has to do when started is to lower the elevator until it detects the ground level. From then on, it can rely on the cab sensor to detect its position.

Now your elevator is able to properly navigate up and down. You have one last problem to solve: How do you inform your elevator which floor it should go to? Placing a touch sensor at every floor to *call* the elevator there is impractical. You have only one input port left on your RCX. What could you do with a single sensor? Can you apply the previous approach here, too?

Yes. You can *count* the pushes on a single touch sensor. For example, three clicks means third floor, and so on. Now you are ready to actually build your elevator!

Bricks & Chips...

Counting Clicks

The following examples are written using a *pseudo-code*—that is, a code that does not correspond to any real programming language, but rather lies between a programming language and natural language. Using pseudo-code is a common practice among professional programmers;

Continued

you are "playing computer" and quickly stepping through an operation in your head to plan and understand what your program will do.

Counting how many times a touch sensor is pressed requires some tricks. Suppose you write some simple code, like this:

```
Counter = 0
repeat
  if Sensor1 is on then
    Counter = Counter+1
  end if
end repeat
```

Your code executes so fast on your RCX that during the short instant you keep the touch sensor pressed, it counts too many clicks. Thus, you need to have it wait for the button to be released before counting a new click:

```
Counter = 0
repeat
  if Sensor1 is on then
    Counter = Counter+1
    wait until Sensor1 is off
  end if
end repeat
```

Now, your code counts properly the transitions from off to on. There's one last feature you must introduce in your code: You want the counting procedure to end when it doesn't receive a click for a while. To do this, you employ a timer that measures the elapsed time from the last click:

```
Counter = 0
Interval = <a proper value>
reset Timer
repeat
  if Sensor1 is on then
    Counter = Counter+1
    wait until Sensor1 is off
```

Continued

```
             or until Timer is greater then Interval
      reset Timer
    end if
  until Timer is greater then Interval
```

Let's say your interval is two seconds. When the counting procedure begins, it resets the timer and the counter to 0 then starts checking the sensor. If nothing happens in two seconds, it exits the **repeat** group. If a click occurs, it counts it, waits for the user to release the button, and resets the timer so the user has again two seconds for another click before the procedure ends.

Light Sensor

Saying that the *light sensor* (Figure 4.5) "sees" is definitely too strong a statement. What it actually does is detect light and measure its intensity. But in spite of its limitations, you can use it for a broad range of applications.

Figure 4.5 The Light Sensor

The most important difference between the touch sensor and the light sensor, is that the latter returns many possible values instead of a simple on/off state. These values depend on the intensity of the light that hits the sensor at the time you read its value, and they are returned in the form of percentages ranging from 0 to 100. The more light, the higher the percentage. What can you do with such a device? A possible application is to build a light-driven robot, a *light follower* as it's called, that looks around to find a strong (or the strongest) light source and directs itself toward it. Provided that the room is dark enough not to produce interference, you could then control your robot using a flashlight.

This ability to trace an external light source is interesting, but probably not the most amazing thing you can do with this sensor. We introduce here another feature of this device: not only does it detect light, but it *emits* some light as well. There is a small red LED that provides a constant source of light, thus allowing you to measure the reflected light that comes back to the sensor.

When you want to measure reflected light, you must be careful to avoid any possible interference from other sources. Remember that this sensor is very sensitive to IR light, too, like the one typically emitted by remote controls, video cameras, or the LEGO IR tower.

Designing & Planning...

Reading Ambient Light

The LEGO light sensor is actually not a great device to measure external sources, as its sensitivity is too low. The emitting red LED is so close to the detector that it strongly influences the readings. If your target is an external source, you might consider trying to reduce the effect of the emitting LED. A simple solution is to place a 1 x 2 one-hole brick just in front of the light sensor. Much more effective solutions require that you slightly modify your sensor. On his Web site, Ralph Hempel shows how to make modifications that neither permanently alter nor damage your sensor (see Appendix A).

The amount of light reflected by a surface depends on many factors, mainly its color, texture, and its distance from the source. A black object reflects less light than a white one, while a black matte surface reflects less light than a black shiny surface. Plus, the greater the distance of the objects from the sensor, the less light returns to the detector.

These factors are interdependent, meaning that with a simple reading from your light sensor, you cannot tell anything about them. But if you keep all the factors constant except one, you are now able to deduce many things from the readings. For example, if your light sensor always faces the same object, or objects with the same texture and color, you can use it to measure its *relative distance*. On the other hand, you can place different objects in front of the sensor, at a constant distance, to recognize their *color* (or, more accurately, their *reflection*).

Measuring Reflected Light

To illustrate the concept of measuring reflected light, let's prepare an experiment. Take your RCX, turn it on, attach a light sensor to any input port, and configure the port properly using the Test Panel of your MINDSTORMS box (the red LED should illuminate). Prepare the environment. You need a dark room, not necessarily completely dark but there should be as little light as possible. The RCX has a *console* mode that allows you to view the value of a sensor in real time. Press the **View** button on your RCX until a short arrow in the display points to the port the sensor is attached to. The main section of the display shows the value your sensor is reading. Now you can proceed. Put the light sensor on the table. Take some LEGO bricks of different colors and place them one by one at short distances from the sensor (about 0.5 in., or 1 to 1.5 cm). Keep all of them separated from each other at the same distance, and look at the readings. You will notice how different colors reflect a different amount of light (you might want to write down the values on a sheet).

For the second part of the experiment, take the white brick and move it slowly toward the sensor and then away from it, always looking at the values in the display. You see how the values decrease when you increase the distance. You can find a distance where the white brick reads the same value you have read for the black one at a shorter distance. This is what we meant to prove: You cannot tell the distance *and* the color at the same time, but if you know that one of the properties doesn't change, you can calculate the other. We stress again that in both cases you must do your best to shield your system from ambient light.

Bricks & Chips…

Understanding Raw Values

Understanding raw values is an advanced topic, and not strictly necessary to successfully using the MINDSTORMS system. That said, it does help to understand how to work with sensors.

The RCX converts the electrical signals coming from sensors (of any type) into whole numbers in the range of 0 to 1023, called *raw values*. When, in your program, you configure a port to host a specific kind of sensor, the RCX automatically scales raw values to a different range,

Continued

suitable for that particular kind of sensor. For example, readings from touch sensors become a simple 1 or 0 digit, meaning on or off, while readings from a temperature sensor convert into Celsius or Fahrenheit degrees. Similarly, light sensor readings are converted into percentages through use of the following equation:

Percentage = 146 - raw value / 7

Why should you need to know about this conversion? Well, for most applications the percentage light value returned by the RCX works well, but there are situations where you need all the possible resolution your sensor can provide, and this conversion into percentages masks some of the resolution your light sensor is capable of. Let's explain this with an example. Suppose that, in two different conditions, your light sensor returns raw values of 707 and 713. Convert these numbers into percentages, considering that RCX uses whole numbers only, and thus rounds the result of a division to the previous integer:

146 - (707 / 7) = 146 - 101 = 45
146 - (713 / 7) = 146 - 101 = 45

The 101 in the second equation should have been 101.857..., but it's been truncated to 101, and you lost the difference between the two readings. We agree that in most situations this granularity of readings is not very important, but there are others where even such a small interval matters.

If you program your RCX using RCX Code, the graphic LEGO environment, you must accept the scaled values, because you have no way to access raw values. But if you use *alternative* programming tools you can choose to receive the unprocessed raw values directly, taking advantage, when necessary, of their finer resolution.

Reading colors is a very common application for light sensors. We already explained that the sensor doesn't actually read colors, rather it reads the reflected light. For this reason, it's hard to tell a black brick from a blue one. But, for now, let's continue to use the expression *reading colors*, now that you know what's really behind the reading.

Line Following

Probably the most widespread usage of the light sensor is to make the robot read lines or marks on the floor where it moves. This is a way to provide artificial landmarks your robot can rely on to navigate its environment. The simplest case is *line following*. The setup for this project is very simple, which is one of the reasons it's so popular. Despite its apparent simplicity, this task deserves a lot of attention and requires careful design and programming. We will discuss this topic in greater detail in Part II; for now, though, we want to bring your attention to what happens when the light sensor "reads" a black line on a light floor.

When the sensor is on the floor, it returns, let's say, 70 percent, while on the black line, it returns 30 percent. If you move it slowly from the floor to the line or vice versa, you notice that the readings don't leap all of a sudden from one to the other, they go through a series of intermediate values. This happens because the sensor doesn't read a single point, but a small area in front of it. So when the sensor is exactly over the borderline, it reads half the floor and half the black strip, returning an intermediate result.

Is this feature useful? Well, sometimes it is, sometimes it's not. When dealing with line following in particular, it is *very* useful. In fact, you can (and should) program your robot to follow the "gray" area along the borderline rather than the actual black line. This way when the robot needs to correct its course, it knows which direction to turn: If it reads too "dark," it should turn toward the "light" region, and vice versa.

Designing & Planning…

Calibrating Readings

Sometimes you can't know in advance what actual values your sensor is going to read. Suppose you're going to attend a line following contest: You cannot be sure of the values your sensor will return for the floor and the black line. In this case, and as a good general practice, it is better not to write the expected values as constants in your program, but allow your robot to read them by itself through a simple calibration procedure. Staying with the line following example, you can dedicate a free input port to a touch sensor to be manually pressed when you put your robot on the floor and then on the black line, so it can store the maximum and minimum readings. Or you can program the robot to perform a short exploration tour to uncover those limits itself.

When you need to navigate a more complex area, one, for example, that includes regions of three different colors, things get more difficult. Imagine a pad divided into three fields: white, black, and gray. How can you tell the gray area from the borderline between the white and the black? You can't, not from a single reading, anyway. You must take into consideration other factors, like previous readings, or you can make your robot turn in place to make it gather more information and understand where it is. To handle a situation like this, your software is required to become much more sophisticated.

The light sensor is such a versatile device that you can imagine many other ways to employ it. You can build a form of proportional control by placing a multicolor movable block of LEGO parts in front of it. Figure 4.6 shows an example of this kind. When you push or pull the upper side of the beam, the sensor reads different light intensities.

Figure 4.6 An Analogue Control with a Light Sensor

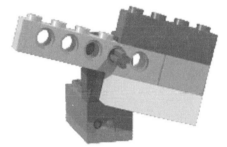

Combining the light sensor with a lamp brick (not included in the MIND-STORMS kit) you get a photoelectric cell (Figure 4.7); your robot can detect when something interrupts the beam from the lamp to the sensor. Notice that we placed a 1 x 2 one-hole beam in front of the light sensor to reduce the possible interference from ambient light.

Figure 4.7 A Photoelectric Cell

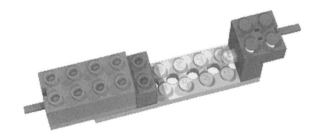

Proximity Detection

You can also use the light sensor as a sort of radar to detect obstacles *before* your robot hits them. This is called *proximity detection*. The technique is based on a property we have already discussed and explored: that the light sensor can be used to measure relative distances based on reflected light. Suppose your robot is going straight, with a light sensor pointing ahead of it. Suppose also that your robot moves in a dark room, with no other sources of light except the emitting red LED of the sensor. While moving forward, the robot continuously reads the sensor. If the readings tend to increase rapidly, you can deduce that the robot is going toward something. There's nothing you can tell about the *nature* of the obstacle and its distance, but if nothing else moves inside the room, you are sure the robot is getting closer to the obstacle. Great! We now have a system to *avoid* obstacles instead of being limited to detecting them through collisions.

NOTE

The red LED in the sensor emits visible light, while the IR LED in the RCX emits invisible light!

Unfortunately, this technique doesn't work very well in a room with any source of light, because your program cannot tell the difference between its red light reflected back, or any other change in the ambient light. You would need a stronger independent source on the robot to provide a better reference. Thankfully, you just happen to have one! The RCX has an IR LED to send messages to the tower or to another RCX. Sending a message uses the IR LED in the RCX to encode the bits in a format that can be received by the tower. We don't care about the contents of the message; we just want the light. Infrared light, though not visible to the human eye, is of the very same nature as visible light, and the LEGO light sensor happens to be very sensitive to it.

So you now have all the elements to use proximity detection in your programs. Send an IR message and immediately read the light sensor. You had better average some readings to minimize the effect of external sources (we'll discuss this trick in Chapter 12). If you notice a significant increase between two subsequent groups of readings, for example, ten percent, your robot is very likely headed towards an obstacle.

Rotation Sensor

The third LEGO sensor we'll examine is the *rotation sensor* (Figure 4.8). It's a pity this piece of hardware is not included in the MINDSTORMS kit, its versatility being second only to the light sensor. However, there is one included in the 3801 Ultimate Accessory Set, together with a touch sensor, a lamp brick, the remote control, and a few other additional parts.

Figure 4.8 The Rotation Sensor

Bricks & Chips...

How the Rotation Sensor Works

The rotation sensor returns four possible values that correspond to four states, let's call them A, B, C, and D. For every complete turn, it passes through the four states four times—that's why we get 16 counts per turn. Turning the sensor clockwise, it will read the sequence ABCDA..., while turning it counterclockwise will result in the sequence ADCBA.... The RCX polls the sensor frequently, and when it detects that the state has changed, it can not only deduce that the sensor has turned, but also tell in which direction it has turned. For example, transitions from A to B, or from D to A, increment the counter by one unit, whereas transitions from D to C, or from A to D, decrement it by one unit.

The rotation sensor, as its name suggests, detects rotations. Its body has a hole that easily fits a LEGO axle. When connected to the RCX, this sensor counts a unit for every sixteenth of a turn the axle makes. Turning in one direction, the count increases, while turning in the other, the count decreases. This count is *relative* to the starting position of the sensor. When you initialize the sensor, its count is set to zero, and you can reset it again in the code, if necessary.

By counting rotations, you can easily measure position and speed. When connected to the wheels of your robot (or to some gearing that moves them), you can deduce the traveled distance from the number of turns and the circumference of the wheel. Then you can convert the distance into speed, if you want, dividing it by the elapsed time. In fact, the basic equation for distance is:

distance = speed x time

from which you get:

speed = distance / time

If you connect the rotation sensor to any axle between the motor and the wheel, you must remember to apply the proper ratio to the count you read. Let's do an example along with the math together. In your robot, the motor is connected to the main wheels with a 1:3 ratio. The rotation sensor is directly connected to the motor, so it shares the same 1:3 ratio with the wheel, meaning that every three turns of the rotation sensor, the wheels make one turn. Every rotation of the sensor counts 16 units, so 16 x 3 = 48 increments, which correspond to a single turn of the wheel. Now, to calculate the traveled distance we need to know the circumference of the wheel. Luckily, most of the LEGO wheels have their diameter marked on the tire. We had chosen the largest spoked wheel, which is 81.6cm in diameter (LEGO uses the metric system), thus its circumference is 81.6 x π ≈ 81.6 x 3.14 = 256.22cm. At this point, you have all the elements: the distance traveled by the wheel results from the increment in the rotation counter divided by 48 and then multiplied by 256. Let's summarize the steps. Calling R the resolution of the rotation sensor (the counts per turn) and G, the gear ratio between it and the wheel, we define I as the increment in the rotation count that corresponds to a turn of the wheel:

I = G x R

In our example G is 3, while R is always 16 for LEGO rotation sensors. Thus, we get:

I = 3 x 16 = 48

On each turn, the wheel covers a distance equal to its circumference, C. You can obtain this from its diameter D by using the formula:

$C = D \times \pi$

Which, in our case, means (with some rounding):

$C = 81.6 \times 3.14 = 256.22$

The final step is about converting the reading of the sensor, S, into the traveled distance, T, with the equation:

$T = S \times C / I$

If your sensor reads, for example, 296, you can calculate the corresponding distance:

$T = 296 \times 256.22 / 48 = 1580$

The distance, T, results with the same unit you use to express the diameter of the wheel.

Actually doing this math in your program, even it is nothing more than a division and a multiplication, requires some care (something we will discuss later in Chapter 12).

Controlling your wheels with rotation sensors provides a different way to detect obstacles, a sort of indirect detection. The principle is quite simple: If the motors are on, but the wheels don't rotate, it means your robot got blocked by an obstacle. This technique is very simple to implement, and very effective; the only requirement being that the driving wheels don't slip on the floor (or don't slip too much), otherwise you cannot detect the obstacle. You can avoid this possible problem by connecting the sensor to an idle wheel, one not powered by a motor but instead dragged by the motion across the floor: If it stops while you're powering the driving wheels, you know your motor has stalled.

There are many other situations where the rotation sensor can prove its value, mainly by way of controlling the position of an arm, head, or other movable parts. Unfortunately, the RCX has some problems in detecting precisely any count when the speed is too low or too high. Actually, this is not the fault of the RCX itself, but its *firmware*, which misses some counts if the speed is outside a specific range. Steve Baker proved through an experiment that 50 to 300 rpm is a safe range, with no counts missed between those values. However, in ranges under 12 rpm or over 1400 rpm, the firmware will *surely* miss some counts. The areas between 12 rpm and 50 rpm, and between 300 rpm and 1400 rpm, are in a gray area where your RCX *might* miss some counts.

This is a small problem, if you consider that you can often gear your sensor up or down to put it in the proper range.

Temperature Sensor

This is the last sensor of the LEGO MINDSTORMS line. It's an optional sensor, not supplied with the MINDSTORMS kit, but it's easy to get through the LEGO online shop or through their Shop-At-Home service. Let's just say that it's a sensor you can definitely live without, even though it can support some funny projects, like a robot that warns you if your drink is getting too warm or too cold.

There are no movable parts, just a small aluminum cylinder that protrudes from the body of the sensor (Figure 4.9). Depending on how you configure it in your code, you can get the temperature values returned in Celsius or Fahrenheit degrees. It can detect temperatures in the range –20°C to 70°C (–4°F to 158°F), but is very slow in changing from one value to the next, so it's not the best device to use if you're looking to detect sudden changes in the temperature.

Figure 4.9 The Temperature Sensor

NOTE

LEGO sensors come in two families: *active* and *passive* sensors. Passive simply means they don't require any electric supply to work. The touch and temperature sensors belong to the passive class, while the light and rotation sensors are members of the active class.

In case you're wondering how active sensors can be powered through the same wire on which the output returns to the RCX, the answer is that a control circuit cycles between supplying power (for about 3 ms) and reading the value (about 0.1 ms).

The equation used to convert raw values from this sensor into temperatures (in C°) is the following:

C° = (785 − raw value) / 8

Celsius degrees translates into Fahrenheit according to the formula:

F° = C° x 9 / 5 + 32

Sensor Tips and Tricks

Sooner or later you will probably find yourself without the proper sensor for a particular project. For instance, you need three touch sensors, but have only two. Or you need a rotation sensor, but don't have any at all. Is there anything you can do? There's no way to turn any sensor into a light sensor or a temperature sensor, but touch and rotation sensors are at some level replaceable.

Another problem we do battle with every time we build with the RCX is the limited number of ports. Later in the book we'll explore some non–LEGO solutions to this problem, but for now we'll talk about some simple cases where you can connect two or more sensors to the same input port.

In the following sections, you'll find some common and well-tested tips that can help.

Emulating a Touch Sensor

Turning a light sensor into a touch sensor is an easy task; you already know the solution. Basically, you build something similar to what we showed in Figure 4.6. In its default state, the sensor has a LEGO brick just in front of it. The pressure on a lever (or beam, axle, plate, and so on) moves a brick of a different color in front of the light sensor, where your software detects the change. Use a belt to keep your assembly in its default position. Try and protect the light sensor as much as possible from external interferences.

Emulating a touch sensor with a rotation sensor is also doable by building a small actuator that rotates the sensor at least a sixteenth of a turn when touched. One of the many possible approaches is shown in Figure 4.10.

Emulating a Rotation Sensor

There's a long list of possible alternatives to the rotation sensor. All the suggested methods are based upon counting single impulses generated by a rotating part.

They all work well, but usually they don't detect the direction of rotation. In many cases this is not a problem, because when coupled with a motor you *know* which direction your sensor is moving.

Figure 4.10 Emulating a Touch Sensor with a Rotation Sensor

The assembly pictured in Figure 4.11 shows an axle with a cam that closes a touch sensor. This is the principle, use either a cam or any other suitable part that, while rotating, periodically pushes the touch sensor. Counting only a single tick per turn, this sensor has a very low resolution. You can increase it by making the sensor close more than once per turn, or simply by gearing the sensor up a bit until you get the required accuracy.

Figure 4.11 Emulating a Rotation Sensor with a Touch Sensor

Making a rotation sensor out of a light sensor is not very different: build some kind of rotating disk with sectors of different color, and count the transitions from one color to the other (Figure 4.12). The general tip for the light

sensor applies to this case, too: Try and insulate the light sensor from external light sources as much as possible.

Figure 4.12 Emulating a Rotation Sensor with a Light Sensor

There are two LEGO electric devices that, though not being actual sensors, can be successfully employed to emulate a rotation sensor. None of them is included in the MINDSTORMS kit, but they're not hard to find.

The first is the polarity switch we introduced in Chapter 3. Connect it as shown in Figure 4.13, and configure it as a touch sensor. With every turn, it closes the circuit twice.

Figure 4.13 Emulating a Rotation Sensor with a Polarity Switch

> **WARNING**
>
> When using a polarity switch to emulate a rotation sensor, due to higher friction, the polarity switch cannot rotate as freely as a true rotation sensor.

The second is the Fiber-Optic System (FOS) device (Figure 4.14). Designed to be mainly a decorative item, this unit, when powered, emits a red light, and by rotating it, you can address the light to one of eight possible small holes. Despite its original purpose, it works surprisingly well as a rotation sensor. Try and connect it directly to an input port of your RCX and configure it as a light sensor; rotate it slowly while viewing its values on the display. They swing from about 70 percent down to 2 percent, then back to 70 percent. You can count sixteen transitions per turn. Thus, the resolution is the same as the original rotation sensor. It has a very low friction quotient, too, resulting in an ideal substitute.

Figure 4.14 The Fiber Optic System Unit

Connecting Multiple Sensors to the Same Port

In some particular cases, connecting multiple sensors to the same port is doable and safe for your devices. Touch sensors, for example, are easy to combine together in an OR configuration, meaning that if any in a group of them gets pressed, you read an *on* state. This is very easy to achieve, simply stacking all the connectors that come from the sensors on the same port. You cannot tell which one was pressed, but there are indeed situations where you can deduce this information from other known facts. For example, say you have a robot with front and rear bumpers. You can connect them to two switches wired to the same port.

When a bumper closes, your program knows if the robot was going forward or backward, thus it can properly interpret the information and behave accordingly. In another example, perhaps your moving robot has a lift arm that requires a limit switch to stop at a specific position. If your robot is stationary when it activates the arm, you can safely use the same port for that limit switch and for a switch wired to a bumper.

The scheme to combine two sensors in an AND configuration is a bit more complex. Tom Schumm came up with the solution shown in Figure 4.15. It works well, provided you connect the wires exactly as shown in the diagram. You get an *on* state only when both sensors are pressed at the same time. The scheme can be extended for an AND configuration using more than two sensors, though it's hard to imagine a situation where you might need such a combination.

Figure 4.15 Connecting Two Touch Sensors in an AND Configuration

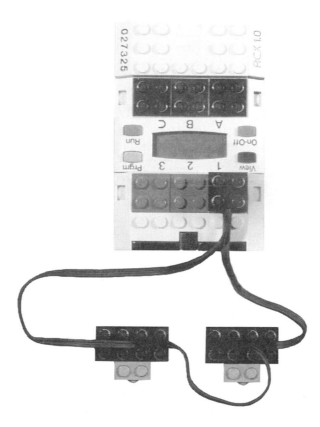

The other sensors (light, rotation, and temperature), don't go together well when used on the same port. If you combine any of them with others of the same or different kind, you will get unpredictable or useless results. There's only one very significant exception to this rule: You can connect a light sensor and a touch sensor to the same port (when configured for a light sensor). This trick, suggested by Brian Stormont, works because the light sensor actually never reads more than about 90 percent, but when the touch sensor gets pressed, the reading jumps to 100 percent, allowing your code to detect the event. The only drawback is that you cannot read the light intensity while the touch sensor is closed. But there are many situations where the touch receives only short impulses, and by applying this trick, you conserve one of your input ports.

Other Sensors

There are other kinds of sensors in the LEGO universe, but we won't discuss them in great detail because they are either difficult to acquire or not very useful. The Cybermaster set (code 8482) includes three touch sensors that are very similar to the MINDSTORMS variety, but that come in three flavors recognizable by the different colors of the buttons (see Figure 4.16). Their transparent casing allows you to see the internal mechanism, which feature internal resistors of different values. For this reason, in terms of raw values, they return different individual readings. This means you can wire them to the same port, and by reading the resulting raw value, determine which one was pressed.

Figure 4.16 The Cybermaster Touch Sensors

The LEGO DACTA line of products includes other sensors designed to survey weather conditions (like humidity) or other specific quantities. They are of no general use, and tend to be very expensive.

Many people have developed their own designs when building custom sensors, and some of them are quite useful if you're open to adding nonoriginal parts to your system. We'll return to this topic in Chapter 9.

If you want to learn more about how LEGO (and non-LEGO) sensors work, don't miss the reference material in Appendix A, and be sure to check out Michael Gasperi's site as well. He is an authority in this field, having discovered many functional details himself, and so displays them in his Web site along with useful information collected from other people.

Summary

In this chapter, we've introduced you to the world of sensors, four basic types in particular: touch, light, rotation, and temperature. Their basic behavior is easy to understand, but here you've discovered that if you want to get the very most out of them, you must study them in greater detail. The touch sensor, for example, seems to be a simple device, but with some clever work on your part, it can become an important tool for counting clicks, or can make a good bumper.

You were also introduced to the light sensor, a small piece of incredibly versatile hardware, which can act as a substitute for both the touch and rotation sensors with minimal effort. Together with the IR LED of the RCX, it makes proximity detection possible, a technique that allows your robot to avoid obstacles before it physically touches them.

The rotation sensor will be your partner in the most sophisticated project of this book. Now you know how it works, and also how to replace it in case you don't have one.

Only the temperature sensor received very little attention. It's the Cinderella of this chapter, basically because it has very limited applications. Nevertheless, it will have its moment of glory at the end of Part II.

Chapter 5

Building Strategies

Solutions in this chapter:

- Locking Layers
- Maximizing Modularity
- Loading the Structure
- Putting It All Together: Chassis, Modularity, and Load

Introduction

Having discussed motors and sensors, and geometry and gearings, it's now time to put all these elements together and start building something more complex. We stress the fact that robotics should involve your own creativity, so we won't give you any general rule or style guide, simply because there aren't any. What you'll find in this short chapter are some tips meant to make your life easier if you want to design robust and modular robots.

Locking Layers

Recall the standard grid we discussed in Chapter 1. We showed how it leads to easy interlocking between horizontal and vertical beams. The sequence was: 1 beam, 2 plates, 1 beam, 2 plates...

You can take advantage of the plate layer between the beams to connect two groups of stacked beams, thus getting a very simple chassis like the one in Figure 5.1. If you actually build it, you can see how, despite its simplicity, it results in a very solid assembly. This also proves what we asserted in Chapter 1 regarding the importance of locking layers of horizontal beams with vertical beams. For instance, if you remove the four 1 x 6 vertical beams, the structure becomes very easy to take apart.

Figure 5.1 A Simple Chassis

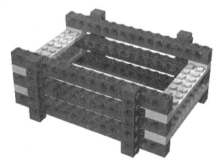

You're not compelled to place all the beams in one direction and the plates in another. Actually, you are likely to need beams in both directions, and Figure 5.2 shows a very robust way to mount them, locked in the intermediate layer of our example structure.

Figure 5.2 Alternating Plates and Beams

NOTE

Remember to use the *black pegs* (or *pins*) when connecting beams. They fit in the holes with much more friction than the gray ones, because they are meant to block beams. The *gray pegs*, on the other hand, were designed for building movable connections, like levers and arms.

Sometimes you want to block your layers with something that stays *inside* the height of the horizontal beams, maybe because you have other plates or beams above or below them. The full beams we've used up to this point extend slightly above and below the structure. The *liftarms* help you in such cases, as shown in the three examples of Figure 5.3:

Liftarm a Two coupled 1 x 5 liftarms with standard black pegs

Liftarm b A single 1 x 5 liftarm and .75 dark gray pegs

Liftarm c Two 1 x 3 liftarms with axle-pegs

NOTE

Naming all the individual LEGO parts is not an easy task. Some people call a *half-beam* what we refer to as a *liftarm*, because it has half the thickness of a beam. Due to this, we chose to use the terminology defined in a widely accepted source: the LUGNET LEGO Parts Reference (see Appendix A for the URL to the site).

Figure 5.3 Using Liftarms to Lock Beams

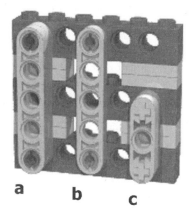

Despite our insistence on the importance of locking beams, there's no need to go beyond the minimum required to keep your assembly together. When the horizontal beams are short, a single vertical beam is usually enough. The example a in Figure 5.4 is better than its b counterpart, because it reaches the same result with fewer parts and less weight. Weight is, actually, a very important factor to keep under control, especially when dealing with mobile robots. The greater the weight, the lower the performance, due to the inertia caused by the mass and because of the resulting friction the main wheel axles must endure.

Figure 5.4 One Vertical Beam Is Sometimes Enough

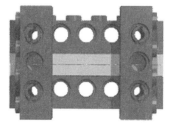

a   b

Bricks & Chips…

What Is Inertia?

In physics, *inertia* is the tendency objects have to resist changes when in states of motion or rest. Objects at rest tend to stay at rest, while objects in motion tend to stay in motion, moving with the same direction and speed. All objects have this tendency, but some more so than others: chiefly because inertia depends on *mass* (quantity of matter). A good example of how mass affects inertia comes from something with which most people have a direct experience: shopping carts! When the cart is empty, you can easily start and stop it, or change its direction, with minimum effort. The more stuff you put inside, however, the more strength is required to maneuver it. Why? Because its objective mass, and thus its inertia, has increased. Similarly, the greater the mass of your robot, the more force is required from its motors when accelerating or braking.

Maximizing Modularity

While building your robot, you will likely have to dismantle and rebuild it, or parts of it at least, many times. This isn't like following someone's detailed instructions; it's more of a trial and error process. Unless you're a very experienced builder and are blessed with clear ideas, your design will develop in both your mind and your hands at the same time.

For this reason, it's best to make your model as easy to take apart as possible, or, to term it more appropriately, your robot should be *modular* in construct. Building in a modular fashion also gives you the opportunity to reuse components in other projects, without having to rebuild common subsystems that already work. This is not always possible, because when you want something really compact, you have to trade away some modularity in favor of tighter integration. Nevertheless, it's a good general building practice, especially when constructing very large robots.

The same principle applies to your most important components: motors, sensors, and obviously the RCX itself. If you are becoming obsessed by LEGO robotics, you will probably buy some extra parts to expand your building possibilities. You might have the resources to start more than one project at a time, or to not be forced to dismantle your last robot when building a new one—however

RCXs, motors, and sensors are definitely not cheap, so your best option is to install them in a way that makes them easy to remove without having to break your robot down into single parts.

NOTE

One good reason to make your RCX easily detachable is that you must be able to change batteries when necessary. The most common solution is to keep the RCX at the very top of your robot—this way you can also easily access the push buttons and read the display.

Loading the Structure

Even the most minimal configuration of a mobile robot has to carry a load of about 300g (11 oz): the weight of one RCX (with batteries) and two motors. Adding cables, sensors, and other structural parts, can easily push you up to about 500g (18 oz). Should you worry about this mass? Is its position relevant?

The first factor you need to consider is friction. You should take all possible precautions to minimize it. This is especially true where the structure attaches to the wheels, because it is there that you transfer all the weight to the wheels by way of the axles. The wheel acts as a lever: the greater the distance from its support, the greater the resulting force on the axle. Such forces tend to bend axles, twist beams, and produce plenty of friction between the axle and the beam itself. For this reason, it's important you keep your wheel as close as possible to its supporting beam. Figure 5.5 shows three examples: a being the worst case, with c the best.

Figure 5.5 Keep a Wheel as Close as Possible to Its Supporting Beam

a - good   b - better   c - best

We suggest you also support the load-bearing axles with more than a single beam whenever possible. The three examples shown in Figure 5.6 are better than those in Figure 5.5, with 5.6c being the best among all the solutions shown so far. The use of two supports, one on either side of the wheel, like on a bicycle, avoids any lever-effect created by the axle on the support, thus reducing the friction to a minimum.

Figure 5.6 Two Supporting Beams Are Better than One

a - good   b - better

c - best

The position of the RCX has a strong influence on the behavior of mobile robots. It's actually the shape and weight of the whole robot that determines how it reacts to motion, but the RCX (with batteries) is by far the heaviest element and thus the most relevant to balancing load. To explain *why* balancing load is important, we must recall the concept of *inertia*. We explained earlier in the chapter that any mass tends to resist a change in motion. In some cases, to resist *acceleration*. The greater the mass, the greater the force needed to achieve a given variation in speed.

The Acrobot model shown in the MINDSTORMS Constructopedia works under this same principle. If you have already built and tried it, did you wonder why it turns upside down instead of moving forward? This happens because the inertia of the robot keeps it in its present condition—which is stationary. Once power is supplied to the motor, the wheels try to convert that power into motion, accelerating the robot. But the inertia is so great that the force resorts to the path with least resistance, turning the body of the robot instead of the wheels. After having turned upside down, the robot has the undriven wheels in front of it, preventing it from turning again, and now can't do anything other than accelerate.

You probably don't want your robots to behave like Acrobot. More likely, you're looking for stable robots that don't lose contact with the ground. You can use gravity to counteract this unwanted effect, putting most of the weight further from the driving axles. There's no need for complex calculations, simply experiment with your robot, running a simple program that starts, stops, reverses, and turns the robot to see what happens. Place the RCX in various positions until you're satisfied with the result.

Putting It All Together: Chassis, Modularity, and Load

The following example summarizes all the concepts discussed so far in this chapter. Using only parts from the MINDSTORMS kit, we built the chassis shown in Figure 5.7. Its apparent simplicity actually conceals some trickiness. Let's explore this together.

Figure 5.7 A Complete Platform

It's built like a sandwich, with two layers of beams that contain a level of plates. It's robust, because vertical beams lock the layers together. Notice that for the inner part of the robot, we used 1 x 3 liftarms instead of 1 x 4 beams. This way the top results in a smooth surface where one can easily place the RCX or other components.

The load-bearing axles are two #8 axles that support both the outer and inner beams (#8 means that the axle is 8 studs long), while the wheels are as close as possible to their supports.

The motors have been mounted with the 1 x 2 plates with rail, as explained in Chapter 3 (look back to Figure 3.4). They are kept in place by two 2 x 4 plates on their bottom (Figure 5.8), but by removing those plates you can quickly and easily take out the motors without altering the structure (Figure 5.9).

Figure 5.8 Bottom View

You can also remove the pivoting wheel and the two main wheels in a matter of seconds to reuse them for another project (Figure 5.10). We should mention here that the pivoting wheel is quite special, since it's what makes a two-wheeled robot stable and capable of smooth turns. The technique of making a good pivoting wheel has its own design challenges, of course, which we'll explore in Chapter 8.

The truth is that if you own only the Robotic Invention System, you probably won't have enough parts to build another robot unless you dismantle the whole structure. If you have more LEGO TECHNIC parts, however, you can leave your platform intact and reuse wheels and motors in a new project.

Figure 5.9 Removing the Motors

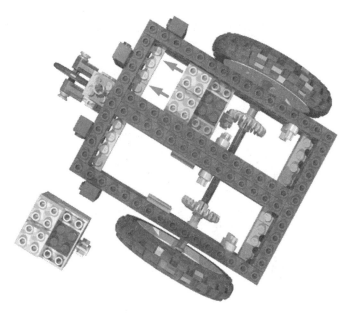

Figure 5.10 ...and the Wheels

Now we can experiment with load and inertia. If you have the LEGO remote control, you don't need to write any code. If not, we suggest you write a very short program that moves and turns the robot. You don't need anything more complex than the following pseudo-code example, which will drive your robot briefly forward then backward, and make it turn in place:

```
start left & right motors forward
wait 2 seconds
stop left & right motors
wait 2 seconds
start left & right motors reverse
wait 2 seconds
stop left & right motors
wait 2 seconds
start left motors forward
start right motors reverse
wait 2 seconds
stop left & right motors
```

Place your RCX in different locations and test what happens. When it is just over the main wheel axles (Figure 5.11), the robots tend to behave like the Acrobot and overturn easily.

Figure 5.11 Poor Positioning of the Load RCX Makes This Robot Unstable

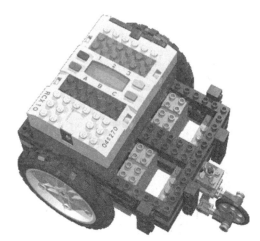

As you move the RCX toward the pivoting wheel, the robot becomes more stable (Figure 5.12). It still jumps a bit on sudden starts and stops, but it doesn't flip over anymore.

Figure 5.12 Better Positioning Improves Stability

Summary

The content of this chapter may be summarized in three words: layering, modularity, and balancing. These are the ingredients for optimal structural results.

Thinking of your robot in terms of *layers* will help you in building solid, well-organized structures. Recall the lessons you learned in Chapter 1 about layering beams and plates and bracing them with vertical beams to get a solid but lightweight structure. A robust chassis comes more from a good design than from using a large number of parts.

Modularity can save you time, allowing you to reuse components for other projects. This is especially important when it comes to the "noble" parts of your MINDSTORMS system—the sensors, motors and, obviously, the RCX—because they are more difficult and expensive to replicate. You should put this concept into operation not only for single parts, but for whole subsystems (for example, a pivoting wheel), which you can transfer from one robot to another.

Balancing is the key to stable vehicles. Keep the overall mass of your mobile robots as low as possible to reduce inertia and its poor effects on stability. Experiment with different placements of the load, mainly in regards to the RCX, to optimize your robot's response to both acceleration and deceleration. We will look more deeply into this matter in Chapter 15, when we learn how to build walking robots (where management of balance is a strict necessity).

Unfortunately, these goals are not always reachable; sometimes other factors force you to compromise. Compactness, for example, doesn't mesh well with modularity. Certain imposed shapes, like those used in the movie-inspired droids of Chapter 18, can force you to bypass some of the rules stated here. We aren't saying they can't be violated. Use them as a guide, but feel free to abandon the main road whenever your imagination tells you to do so.

Chapter 6

Programming the RCX

Solutions in this chapter:

- **What Is the RCX?**
- **Using LEGO RCX Code**
- **Using the NQC Language**
- **Using Other Programming Languages**
- **Divide and Conquer: Keeping Your Code Organized**
- **Running Independent Tasks**

Introduction

As we explained in the Introduction, this book is not about programming—there are already many good resources about programming languages and techniques, and about programming the RCX in particular. However, the nature of robotics (often called *mechatronics*) is such that it combines the disciplines of mechanics, electronics, and software, meaning you cannot discuss a robot's mechanics without getting into the software that controls the electronics that drives the machine. Similarly, you cannot write the program without having a general blueprint of the robot itself in your mind. This applies to the robots of this book as well. Even though we are going to talk mainly about building techniques, some projects have such a strong relationship between hardware and software that explaining the first while ignoring the latter will result in a relatively poor description. For these reasons, we cannot simply skip the topic, we need to lay the foundations that allow you to understand the few code examples contained in the book.

In the previous chapters, we mentioned the RCX many times, having assumed that you are familiar with the documentation included in the MIND-STORMS kit and know what the RCX is. The time has come to have a closer look at its features and discover how to get the most from it. We will describe its architecture and then give you a taste of the broad range of languages and programming environments available, from which you can choose your favorite. Our focus will be on two of them in particular: RCX Code, the graphic programming system supplied with the kit, and NQC, the most widespread independent language for the RCX.

The last sections of the chapter provide a complete code example, which is meant to help explain how to write well-organized code that is easy to understand and maintain, and is designed to familiarize you with the programming structures you'll find later in the book.

What Is the RCX?

The RCX is a computer. You are used to seeing computers that have a keyboard, a mouse, and a monitor—devices created to allow human users to interface with their computers—but the RCX hasn't got any of those features. Its only gates to the external world are a small display, three input ports, three output ports, four push-buttons, and an infrared (IR) serial communication interface. The RCX is actually more similar to industrial computers created to control machinery than it is to your normal desktop computer. So, how can you program it if it hasn't any user interface? You write a program on your PC, then transfer it to the RCX with the help of the IR tower (a device designed to work as a link between the PC and the RCX), and, finally, the RCX executes it.

To understand how the RCX works, imagine a structure made of multiple layers. At the very bottom is the processor, an Hitachi H8300, which executes the machine code instructions. The processor cooperates with additional components that convert signals from the ports into digital data, using chips that provide memory for data and program storage. Just as with most computers, the memory of the RCX is made up of two types: read-only memory (ROM) and random access memory (RAM). The content of the ROM cannot be altered or cancelled in any way, since it is permanently written on the chips, while the data in the RAM can be replaced or modified. The RAM requires a continuous power supply in order to retain its content. When the supply breaks, everything gets erased.

Above the processor and circuit layer you find the ROM code. When you unpack your brand new RCX, there's already some code stored in its internal ROM that's aimed at providing some basic functionality to the RCX: input ports signal conversion, display and output ports control, and IR communication. If you are familiar with the architecture of a personal computer, you can compare this ROM code to the basic input/output system (BIOS), the low-level machine code which is in charge of booting the computer at startup and interfacing with the peripherals.

An RCX with just the ROM code is as useless as a personal computer with just the BIOS. On top of the ROM code layer the RCX runs the *firmware*, which, to continue with our comparison to computers, is its *operating system*. The term *firmware* denotes a kind of software the user normally doesn't alter or change in any way; it's part of the system and provides standard functionality, as operating systems do. In RCX, the firmware is not burned into the system like the ROM code, rather it is stored in the internal RAM, and you download it from your PC using the infrared interface. The LEGO firmware was copied to your PC during the installation of the MINDSTORMS CD-ROM, and transferred to your RCX by the setup process.

The firmware is not the final layer of the system: on top of it there's your own code and data. They will be stored in the same RAM where the firmware is, but from a logical standpoint they are considered to be placed at a higher level. As we explained earlier, you write your code on the PC, then send it to the RCX through the infrared interface. The MINDSTORMS software on the PC side, called *RCX Code*, translates your program (made of graphical code blocks) into a compact form called *bytecode*. The RCX receives this bytecode via the IR interface and stores it in its RAM. When you press the **Run** button, the firmware starts *interpreting* the bytecode and converting its instructions into actions.

WARNING

Because the firmware is stored in RAM, it will vanish if your RCX remains without power for more than a few seconds, and you will have to reload it before using your RCX again. When you power off your RCX, the RAM remains supplied just to keep the firmware in existence, and this is the reason why the RCX will slowly drain the batteries even when switched off. If you plan not to use it for more than a few days, we suggest you remove the batteries to preserve them. Remember that when you need your RCX again, you will have to reload the firmware.

Let's summarize the process from the top to the bottom level:

- You write your program using RCX Code, the MINDSTORMS software on the PC side.

- RCX Code automatically translates your program into a compact format called *bytecode*.

- Using the IR link between the PC—via the IR tower—to the RCX, you transfer the bytecode version of your program to the RAM of the RCX.

- The firmware interprets your bytecode and converts it into machine code instructions, calling the ROM code routines to perform standard system operations.

- The RCX processor executes the machine code.

Most of these steps are hidden to the user, who simply prepares the program on the PC, downloads it to the RCX, presses the **Run** button, and watches the program execute.

A Small Family of Programmable Bricks

The RCX belongs to a small LEGO family of *programmable bricks*. The first to appear on the scene was the Cybermaster, a unit that incorporates two motors, three input ports, and one output port. It shares with the MINDSTORMS devices the ability to be programmed from a PC, with which it communicates through the "tower," which in this case is based on radio frequency instead of infrared transmission. But the similarities end here, and the Cybermaster has more limitations than the RCX:

- Its three input ports work with passive sensors only.

- The firmware is in ROM instead of RAM. This means that it's not possible to upgrade it to a newer version.

- The RAM is much smaller than the one in the RCX and can host only very short programs.

The *Scout*, contained in the Robotics Discovery Set, is programmable from the PC with the same IR tower of the RCX (not included in the set), but features a larger display that allows some limited programming, or better said, it allows you to choose from among various predefined behaviors. It features two output ports, two input ports (passive sensors only), and one embedded light sensor. Like for the Cybermaster, the firmware is in ROM and cannot be upgraded or modified.

Using LEGO RCX Code

RCX Code is the graphical programming tool that LEGO supplies to program the RCX. If you have installed the MINDSTORMS CD-ROM, followed the lessons, and tried some projects, you're probably already familiar with it.

RCX Code has been targeted to kids and adults with no programming experience, and for this reason it is very easy to use. You write a program simply by dragging and connecting *code blocks* into a sequence of instructions, more or less like using actual LEGO bricks.

There are different kinds of code blocks that correspond to different functions: You can control motors, watch sensors, introduce delays, play sounds, and direct the flow of your code according to the state of sensors, timers, and counters. RCX Code also provides a simple way to organize your code into *subroutines*, groups of instructions that you can call from your main program as if they were a single code block.

When you think your code is ready to be tested, you download it to the RCX through the IR tower. The RCX has five *program slots* that can host five independent programs. When downloading the code, you choose which slot to download to, and with the **Prgm** push-button on your RCX, you select which program to execute.

The intuitiveness of RCX Code makes it the ideal companion for inexperienced users, but it has some major drawbacks:

- Its set of instructions is very limited, and doesn't disclose all the power your RCX is capable of. Sooner or later you will start desiring a more powerful language.

- Its graphical interface is not suitable for large programs. The sequence of code blocks, though very intuitive for small programs, becomes hard to follow when you have tenths or hundredths of them.

For these reasons, you'll find that RCX Code is a barrier to the development of complex projects.

Using the NQC Language

The LEGO firmware is a solid, well-tested software that provides a rather complete functionality. The surprising thing is that it actually offers many more possibilities than what the RCX Code discloses to us. It's like having a car whose motor is capable of 100 mph but with an accelerator pedal that allows you to reach no more than 50 mph. The power is there, but the interface doesn't allow you to get at it. This fact drove some independent developers to create new programming environments able to get the most from the LEGO firmware, providing access to those features that RCX Code conceals. All of them share the same approach, which consists of making a new interface on the PC side that's able to generate bytecode and transfer it to the RCX.

Developed and maintained by Dave Baum, the language called Not Quite C (NQC) has achieved enormous popularity among MINDSTORMS fans and is by far the most widespread of this category. NQC is based on C-like syntax; if you're not a programmer, or if you have no experience with C, don't be frightened by this. NQC has a very smooth learning curve, and comes with a lot of documentation and tutorials. The success of NQC has come about for many reasons:

- It's based on the original LEGO firmware, thus taking advantage of its ability to produce very reliable code, and at the same time freeing all of RCX Code's hidden power. Even from its very first releases NQC has proven to be rock solid.

- Dave Baum puts a lot of effort into maintaining it, continuously adding new features and acknowledging new opportunities offered by the

LEGO firmware. NQC supported the new RCX 2 firmware version well before it was officially released in any LEGO product.

- It is multiplatform, both on the host side (it runs on PC, Mac, and Linux machines) and on the target side (it supports all the LEGO programmable bricks: RCX, Scout, Cybermaster).

- It is self-contained. To use NQC you don't need any other tools than a simple text editor (Windows Notepad is enough). The installation procedure is as easy as copying a file.

- There are many documents and tutorials, in many different languages, that help new users understand all the details.

- The NQC compiler is a command-line tool, with no user interface, but people have developed nice integrated development environments that encapsulate NQC inside a productive system that includes editors, tools, diagnostics, data logging, and other utilities, as well as, most importantly, the Bricx Command Center.

- NQC is free software released under the Mozilla Public License (MPL).

Some of the projects discussed in this book actually require that you go beyond the limits imposed by RCX Code. This is the main reason why we chose NQC to illustrate the few programming examples. NQC also has the advantage that, being a textual language, it allows for a very compact representation that better suits the format of a book.

Using Other Programming Languages

The fact that LEGO placed the firmware of the RCX in the RAM left the system open to other languages that follow a more radical approach. Instead of substituting the software that produces bytecode on the PC side, they *replaced* the firmware on the RCX. It's important to note that installing any of these alternative environments doesn't entail any risk at all for your RCX. You can always return to your original system.

All the work that has been done in this direction heavily relies on Kekoa Proudfoot's pioneering hacking of the RCX. Kekoa patiently disassembled the LEGO firmware and documented all the routines and their calls, thus laying the foundations for the subsequent alternative firmware versions.

Using legOS

In 1999, Markus Noga started The legOS Project, the first attempt to write a replacement firmware for the RCX. Noga's goal was to bypass all the limitations of the bytecode interpreter to run the code directly on the Hitachi H8300 processor of the RCX. A legOS program is a collection of system management routines that you link to your own C or C++ code and load to the RCX in place of the firmware.

What was initially an individual effort turned into a collective open source project under the Mozilla Public License. The legOS Project is now managed by Luis Villa and Paolo Masetti and maintained by a team of a dozen developers.

The installation is not always straightforward, especially for Windows machines. You need to be a programming expert, because what you have to deal with here is true C, not the simplified and friendly NQC version. You have to manage cross-compilers and Unix emulators if you don't run a Unix-like machine, so legOS is definitely not for everyone. But for this price it unleashes the full power of your RCX up to its last bit. You get full control of any resource and any device, can use any C construct and structure, and can address any single byte of memory. Plus, your code runs at an astonishing speed when compared to the interpreted bytecode.

Using pbForth

The pbForth language (the name stands for *programmable brick FORTH*) is the result of Ralph Hempel's experience in designing and programming embedded systems, a field where FORTH is particularly well suited. Conceived in the sixties, the FORTH language has a strong tradition in robotics, automation, and scientific applications. More than a language, FORTH is an interactive environment. The traditional concepts of editing source files, compiling, linking, and so on, don't translate very well to FORTH; it's mainly a stand-alone system.

Ralph Hempel's implementations make no exception to this rule. You download the pbForth kernel to your RCX, and from that moment on you *dialog* with it using a simple terminal emulator. For this reason, pbForth is very portable and very easy to install on any platform.

If you haven't any experience with FORTH, it will probably seem a bit strange to you in the beginning. The language is based on the *postfix notation*, also called reverse polish notation (RPN), which requires you to write the operator after the operands.

If you decide to give pbForth a try, you will discover the benefits of an extensible system that naturally leads you to program in terms of layers. You might find it challenging to learn, but it's a productive—and fun—tool with which you can write compact and efficient code.

Using leJOS

Jose Solorzano started the TinyVM Project, a small Java footprint for the RCX. TinyVM was designed to be as compact as possible, and for this reason lacked much of the extended functionality typical of Java systems. Over the foundation of TinyVM, Jose and other developers designed leJOS, a fully functional Java implementation that includes floating point support, mathematical functions, multiprogram downloading, and much more. LeJOS is an Open Source project and, like legOS, is under continuous development.

leJOS is the newcomer on the scene of MINDSTORMS programming, but we foresee a great future for it. It's complete, portable (currently to PC and Unix-like machines), very easy to install, fast, efficient, and based upon a widespread language. There are also some visual interfaces under development that will make this system even more attractive to potential users.

Using Other Programming Tools and Environments

We know we didn't cover all the available programming tools for the RCX. There are others, like Gordon's Brick Programmer, or Brick Command, that follow the same solution of NQC and convert a textual program into bytecode. There are also a few more replacements for the firmware, like QC or TinyVM. And, finally, some other tools, like ADA for the RCX, that translate source code into NQC code. They are good tools, solid and well-tested, but we choose to describe the most representative and widespread in each class. We recommend you look at Appendix A for further information about the software we introduced here and about other possible choices; the list is so long we are sure you'll find the tool that fits your needs. In the same appendix, you will find some links to other tools that, though not intended for programming, can help you monitor your RCX, transfer data to the PC, graph the status of the input ports, and more.

Divide and Conquer: Keeping Your Code Organized

Up to this point the few programming examples you met were written in a sort of pseudo-code very close to plain natural language. The use of pseudo-code allows the programmer to "play computer" and understand what the program does, but to complete the projects of the book, some of which are a bit complex, you need a real environment to run and test the code with. We chose to write all the examples using NQC because it combines power with compactness, it's easy to install and learn, and has become a widespread standard among thousands of MINDSTORMS programmers. In the following example, we will describe some of the most important features of NQC, but we strongly recommend you read the documentation available from its official Web site, listed in Appendix A. Even if you don't choose NQC, we're sure you can easily translate our examples into your favorite programming language.

What we said in Chapter 5 about keeping your construction designs modular applies to programming as well. Organizing the code into logical sub units is a good programming practice that will often help you in the debugging process. Unless your robot is designed for a very simple task, try to split its code into blocks that correspond to the different situations it's expected to manage and to the actions it should perform. The Latin motto "divide et impera" applies well to programs: the more you divide the code into small sections, the better you can control and understand the program's behavior.

We will use an example to clarify this concept and introduce other tips: Say your robot has been designed to follow a black line, detect small obstacles with a bumper and remove them from its path by pushing the obstacles away with some kind of arm. As we explained earlier, it's impossible to write a program without having a precise idea of how the robot is designed and what it is expected to do. For the example we are going to illustrate, we made the following assumptions about the robot and the environment:

- The line is darker than the floor.

- The robot will follow the left border of the line (e.g., It turns right to go toward the line, left to go away from line).

- Output ports A and C control the left and right drive wheels respectively.

- Output port B operates the arm.

- Input port 1 is attached to a touch sensor connected to the bumper. It closes (goes from 0 to 1) when the bumper is pressed.

- Input port 2 is attached to a face-down light sensor that reads the line.

Here is the initial code you should write:

```
int floor,line;

task Main()
{
  Initialize();
  Calibrate();
  Go_Straight();

  while(true)
  {
    Check_Bumper();
    Follow_Line();
  }
}
```

The main level of your program is quite simple, because at this point you're not concerned with what **Go_Straight** or the other subroutines mean in terms of actions, you're only concerned with the logic that connects the different situations. You are deciding the rules that affect the general behavior of the robot and you don't want to enter into the details of *how* it can actually go straight. This result is achieved by encapsulating the instructions that make your robot go straight into a *subroutine*, a small unit which "knows" what the robot requires in order to go straight. This approach has another important advantage: Your code will be more general because it doesn't depend on the architecture of the robot. For example, for one specific robot "go straight" will mean switching motors A and C on in the forward direction, while for another it might mean switching on motor B in the reverse direction. When you want to adapt the program to a different architecture, you simply change the implementation details contained in the low-level subroutines, without having to intervene on the logic flow.

Let's come back to your main task to examine it in deeper detail. The first instruction is actually placed before the beginning of the task: It declares that you are going to use two *variables* named *floor* and *line* and intended to contain integer numbers. A variable is like a box with a name written on it: You can place something inside, a specific number—that is, you can *assign* a value to the variable. Or you can watch what's inside the box, *reading* the variable. At this stage, you are neither assigning nor reading the variables, you are simply declaring that you need two of them. In other words, you are asking NQC to prepare two boxes with the names just mentioned.

When the user presses the **Run** button on the RCX, the main task begins. After it has completed initialization and calibration procedures, the program starts the robot in straight motion, then it enters an endless loop where the program continuously manages its two tasks: removing obstacles and following the line. The **while(true)** statement repeats all the instructions delimited by the open and close brace forever. In your case, it will execute the Check_Bumper subroutine, then the Follow_line, then the Check_Bumper again in a continuous loop that only the user can interrupt using the **Run** button.

Everything is clear and simple, as it should be. Now let's have a look at what happens at a lower level in our subroutines.

Any program will typically include an *initialization* section, where you set the motor power, configure the sensors, reset timers and counters and initialize variables. This is not required when you use RCX Code, because it automatically configures the input ports for you. NQC, like the other textual environments, requires that you explicitly declare what kind of sensor you connect to each port:

```
void Initialize()
{
   SetSensor(SENSOR_1,SENSOR_TOUCH);
   SetSensor(SENSOR_2,SENSOR_LIGHT);
}
```

The word **void** is what tells NQC that you are describing a subroutine, and it's followed by the name you choose for it. The **SetSensor** statements are used to configure input port 1 for a touch sensor and input port 2 for a light sensor.

The *calibration* routine is designed to inform your robot of the actual light readings it should expect on its path. We discussed this topic briefly in Chapter 4, explaining that keeping your program independent from particular cases is a good general programming practice. In this example, it means you should not write the light sensor thresholds into the code, but rather give your robot the possibility to read them from the environment, and this is what you have declared the *floor* and *line* variables for.

```
void Calibrate()
{
  WaitBumperPress();
  floor=SENSOR_2;
  WaitBumperPress();
  line=SENSOR_2;
  WaitBumperPress();
}


void Wait_Bumper_Press()
{
  PlaySound(SOUND_DOUBLE_BEEP);
  while (SENSOR_1==0);   // wait for bumper press
  while (SENSOR_1==1);   // wait for bumper release
}
```

This code shows that in some situations you can recycle a sensor and use it for more than a single purpose: during the calibration process, the bumper is used as a trigger to tell the robot that it's time to read a value. It also shows that subroutines can be *nested*. In other words, you can make a subroutine call another subroutine. In this particular case, the **WaitBumperPress** is a small service subroutine that produces a beep and then waits until the bumper switch gets pressed and released.

When you run the program, the calibration procedure begins and informs you with a beep that it waits for the first reading. You place your robot with the light sensor on the floor, far from the line, and push the bumper. The program reads the light sensor and stores that value as a typical "floor" value in the *floor* variable. Then it beeps again while waiting to read the line. You place the robot with the sensor just over the line and push the bumper again, making it detect the "line" light value and store it in the *line* variable. The robot finally beeps again, meaning the calibration process has finished and that the next push on the bumper will put it in motion.

This sort of pre-run phase is quite useful in many other situations, such as when you need to prepare the robot for operations by either reading some environmental variable or resetting mechanisms that might have been left in an unknown state by previous executions.

The **Check_Bumper** procedure is in charge of testing whether the robot has hit an obstacle, and if so, how it should react:

```
void Check_Bumper()
{
  if (SENSOR_1==1)
  {
    Stop();
    Remove_Obstacle();
    Go_Straight();
  }
}
```

It checks the bumper, and, if found closed, stops the robot, calls the **Remove_Obstacle** subroutine to clear the path and then resumes motion. Testing the bumper is as simple as checking if SENSOR_1 has become equal to 1, which means that the touch sensor connected to port 1 has been pressed. You notice that we apply here the same concepts used at the main level: encapsulating details into routines at a lower level.

The **Follow_Line** routine is what keeps your robot close to the line edge— let's say the left edge. If the light sensors read too much of the "floor" value, it turns right toward the line. If, on the contrary, it reads too much of the "line" value, it turns left, away from the line. (See Chapter 4 for a discussion of this method.)

```
void Follow_Line()
{
  #define SENSITIVITY 5
  if (SENSOR_2<=floor+SENSITIVITY)        // reading too "floor"
    Turn_Right();
  else if (SENSOR_2>=line-SENSITIVITY)   // reading too "line"
    Turn_Left();
  else
    Go_Straight();
}
```

The method used in this subroutine deserves some explanation. First of all, the word **#define** tells NQC that the following word denotes a *constant*; for the sake of simplicity, you can consider a constant to be like a variable whose value cannot be changed by the program. In this particular case, your program defines the constant *SENSITIVITY* with the value 5. This value is used together with the *floor* and *line* variables to decide what the robot should do. An example with actual numbers can make the things clearer: suppose the **Calibrate** routine placed the value 55 in the *floor* variable and the value 75 in the *line* variable. The program tests if SENSOR_1 is less than or equal to *floor* + *SENSITIVITY*, which results in 55 + 5 = 60, to decide if the robot has to turn right toward the line. Similarly, it tests if SENSOR_1 is greater than or equal to *floor* − *SENSITIVITY*, which corresponds to 75 − 5 = 70, and if this is the case, it makes the robot turn left, away from the line. While the readings remain greater than 60 and lower than 70, the robot goes straight. You can change the value of *SENSITIVITY* to make your robot more or less reactive to readings: An increase will narrow the range of values that allow the robot to go straight, thus your robot will make more corrections in order to remain close to the edge of the line.

The code you wrote so far is rather general and could work for a broad class of robots. Now the time has come to write the part of the program that depends on the physical architecture of your robot.

The **Go_Straight** routine will be very straightforward in most cases. You know from the initial assumptions that the robot has two side wheels (or tracks) driven by two independent motors. In Chapter 8, we will explore this configuration, called *differential drive*, in greater detail. For the moment, let's stick to the fact that if both the motors go forward, the robot goes forward and straight. If one of the motors stops, the robot turns toward the side of the stationary wheel. This knowledge is enough to write the following routines, which control motion:

```
void Go_Straight()
{
  OnFwd(OUT_A+OUT_C);
}


void Stop()
{
  Off(OUT_A+OUT_C);
}


void Turn_Left()
```

```
{
  Off(OUT_A);

  OnFwd(OUT_C);
}


void Turn_Right()
{
  Off(OUT_C);

  OnFwd(OUT_A);
}
```

Designing & Planning…

Benefits of Designing Modular Code

If you follow the principles illustrated in this chapter when writing a modular and well-structured code, your program will result in greater readability, reusability, and testability:

- **Readability** The program is organized into small sections that are easy to comprehend with just a quick glance. This means that your program will be easier to maintain, and more easily understood by the friends with whom you share it.

- **Reusability** Separating the logic of the program from the instruction related to the physical structure of the robot, you make your code more flexible and reusable for different architectures. The general principle is: the upper levels of the code reflect *what* the robot does, while the lower ones reflect *how* the robot does it.

- **Testability** A nice side effect of well-structured code is that it speeds up your testing procedures, segmenting possible problems into small portions of code. Remove (or comment out) the call to **Follow_Line** from inside the repeat block in the main task: Your robot should simply go straight until it hits an obstacle, then activate the arm and remove it. Conversely, you can remove the call to **Check_Bumper** to turn your robot into a simple line follower!

There's one last routine left: **Remove_Obstacle**. Let's say your robot features a very simple arm that works with a single motor and only requires a timed activation:

```
void Remove_Obstacle()
{
  OnFwd(OUT_B);
  Wait(200);
  OnRev(OUT_B);
  Wait(200);
  Off(OUT_B);
}
```

The statement **Wait(200)** makes the program wait for 200 hundredths of a second, or two seconds. This parameter depends on the time your mechanism needs to remove the obstacle, and it is once again related to the physical structure of the robot.

Your program is now finished and ready to be tested. We hope this example made you realize the benefits of a modular and well-structured code.

Running Independent Tasks

All the tools you can choose from to program your RCX support some form of *multitasking*, that is, they support two or more independent tasks that run at the same time. This is not particularly evident when you use RCX Code, but it's a well-documented feature in all the alternative environments.

Multitasking can be helpful in many situations and it's often a tempting approach, but you should use it with a lot of care because it will not always make your life easier. Let's go back for a moment to our previous example: would multitasking have been a good choice? Didn't your robot have two different tasks to manage: line following and obstacle detection? Well, it did, but they were mutually exclusive—after all, your robot was not following the line *while* it removed the obstacle. In cases like this, and in many others, your robot is asked to perform different activities *one at a time* more often than it is asked to perform different activities *at the same time*. Using multitasking, you would have made your code more complex, because of the additional instructions needed to synchronize the tasks. When the **Remove_Obstacle** task stops the robot, it should communicate the **Follow_Line** task to suspend line following, and communicate again when it can be resumed.

In designing a multitasking application, you are required to move from a sequential, step-by-step flow to an *event-driven* scheme, which usually requires additional work to keep the processes coordinated. While sequential programming is like following a recipe to cook something, you can compare multitasking to preparing two or more recipes at the same time. This is quite a common practice in any kitchen, but requires some experience to manage the allocation of resources (stoves, oven, mixer, blender...), respond to the events (something's ready to be taken out of the oven) and coordinate the operations so the tasks don't conflict with each other. You have to think in terms of *priorities*: Which dish should you put in the oven first? Programming independent tasks implies the same concerns: You must handle the situations where two tasks want to control the same motor or play two different sounds. The RCX is well-equipped to manage resource allocation and to support event-driven programs, and NQC gives you full access to these features. However, most of the effort is still on your shoulders: no tool makes up for the disadvantages inherent in a bad design.

In our experience with LEGO robotics, there are few actual situations where multitasking is absolutely necessary, or even useful. Our suggestion is that you approach it only when your robot performs some really independent activities, like playing background music while navigating a room, or responding to messages while looking for a light source.

Summary

In this chapter, you took some first steps on your path to programming LEGO robots. We started describing the RCX, the LEGO programmable unit that's the core of your robots, to unveil some of its secrets. You discovered how its architecture can be easily understood in terms of layers: your program, its translation into bytecode, the interpreter in the firmware, and the processor which executes the operations.

To create your program on a PC, you can choose from many available tools; we briefly described RCX Code, the original LEGO graphic programming environment, and NQC, the most widely accepted independent language for the RCX. We also reviewed a few other environments—legOS, pbFORTH, leJOS—which follow a more radical approach to the goal of getting the most from the RCX: replacing its firmware.

The second part of the chapter does for programming what the previous chapter did for building: it establishes some guidelines. Oddly enough, the two arenas share a lot, since layered architecture and modularity principles apply just as much to the body of the robot as they do to its brain—with the notable difference that sometimes you have good reason not to follow those principles in the hardware. In other words, there is no excuse for badly organized software! We used a short but complete program written in NQC to put these principles into practice, showing how they can improve the readability, reusability, and testability of your code.

Chapter 7

Playing Sounds and Music

Introduction

The RCX features an internal speaker and the hardware necessary to drive it, thus making your robot able to produce sounds. Do not underutilize this feature! It not only offers you a fun way to give your robots a more defined personality, but gives you a simple communication protocol which will help in testing and debugging your programs.

This is why we decided to devote a book chapter to playing sounds and music with the RCX, even though the topic is more related to programming than to building techniques. However, as we explained in Chapter 6, when you are dealing with robotics, the two matters are seldom separable. For some of the robots described in the second part of the book, sounds are an important component in their interface with the external world; for others, sounds are an interesting addition that enriches their behavior.

If you are not familiar with musical terminology or audio file formats, you might find topics in this chapter a bit complex. But the prize is worth the effort, because the techniques explained here open exciting new opportunities in your robot world. You will discover how to use simple tones, how to write melodies, even how to convert digital audio files into sound effects that can be incorporated into your program!

Communicating through Tones

As we explained in the introduction, the RCX features an internal speaker. There is little evidence of it on the outside: The RCX 1.5 has two very small slits on the sides with the LEGO logo stamped on it, from which the sound emanates. The sound system of the RCX is designed to be accessed from your program; you are not allowed to alter the volume of the speaker, which is predefined, but you have full control over the frequency (pitch) and the duration of the notes. The language Not Quite C (NQC), which we will be using in our examples, includes two basic instructions on how to produce sounds, called **PlaySound** and **PlayTone**. Through the **PlaySound** command, the RCX can output one of six predefined sound patterns, such as a short click, a double beep, or a short sequence of tones:

```
PlaySound(SOUND_CLICK);

PlaySound(SOUND_DOUBLE_BEEP);

PlaySound(SOUND_UP);

PlaySound(SOUND_DOWN);
```

The **PlayTone** command plays a single note of a given pitch (in Hertz) and duration (in hundredths of a second). The following statement plays a tone of 262 Hertz for half a second:

```
PlayTone(262,50);
```

The RCX is capable of reproducing any frequency from 31 Hertz to more than 16,000 Hertz; however, you will usually limit yourself to the frequencies which correspond to the musical notes (see the table in Appendix C). All the programming languages built over the LEGO firmware offer this same feature, while most of the others include some kind of more or less sophisticated control over sound.

Sounds are the most immediate way your RCX has to inform you about a specific situation. There is, of course, the *display*, but it's not always in sight, especially when your robot is running across the room! There's also the *datalog*, the feature that allows your PC to read values accumulated in a special memory area in the RCX, but to use it you must be sitting in front of your computer the whole time. Sounds, on the other hand, can be emitted by the robot without interrupting any other activities, and can be heard by you even if the robot is out of sight or far away.

Through simple sound patterns you can make your robot inform you that an operation has ended, something has gone wrong, its batteries are low, and much more. It can acknowledge the push of a button, or tell you it's waiting for specific input from you, as in the case of the **Calibration** routine described in Chapter 6. At the 1999 Mindfest gathering of MINDSTORMS fans and professionals at the Massachusetts Institute of Technology (MIT), we built a Tic-Tac-Toe-playing robot—a version of which you'll see in Chapter 20—that used different musical themes to inform its human opponent about the result of the game.

Playing Music

Sometimes a sound pattern can give your creatures a specific character. Could you imagine a *silent* reproduction of the famous R2-D2 droid from the Star Wars saga?

Music can enrich the personality of your robot even more then tone sequences. A wrestling robot probably appears more resolute if, while facing its opponents, it plays Wagner's "Ride of the Valkyries" rather than a Chopin piano sonata or nothing at all. Our LEGO reproduction of Johnny Five from the movie Short Circuit—described in Chapter 18—plays the Saturday Night Fever theme song while dancing—but if you switch off the soundtrack, it becomes simply a robot that moves around swinging its arms and head.

Playing music requires that you patiently code every single note into your program. LEGO RCX Code is not a suitable tool for melodies longer than just a few notes, but with other textual languages, like NQC, you can write and store very long songs.

Every note in the song requires two attributes: *pitch* and *duration*—the first expressed by a frequency and the second by a time. You must introduce delays between the notes to let the CPU wait out the note's duration before playing the next note.

```
PlayTone(440,50);
Wait(50);
PlayTone(220,100);
Wait(100);
```

In this example, the RCX plays an A (440 Hertz) that's half a second long, waits for the note to finish, then plays another A (220 Hertz) one octave below the previous note for one second.

The RCX is limited to playing a single note at a time, thus we say it's a *monophonic* device. There's no chance to play chords, which require two or more notes played at the same time, but you can adjust note timing to get various effects. In our previous example, the duration of the first note filled the entire interval before the second note, thus producing a *legato* effect. You can just as easily get a *staccato* effect—shortening the duration of the note inside the interval produced by the **Wait** statement—by introducing a pause with no sound between the two notes:

```
PlayTone(440,10);
Wait(50);
PlayTone(220,100);
Wait(100);
```

Coding a melody by hand is a long and tedious task. What happens if when you're finished you discover that the execution is faster or slower than what you intended? Unfortunately, you'd have to go back and change all the time intervals. A better approach takes advantage of a feature that all textual programming environments offer: the definition of *constants*. Using constants you can make all the intervals relative to a specific duration that controls the execution speed:

```
#define BEAT 50
PlayTone(440, BEAT);
Wait(BEAT);
PlayTone(220, 2*BEAT);
Wait(2*BEAT);
```

This code behaves exactly like our first example, but you'll see that by having defined a constant, the code is clearer and easier to maintain, simply changing the value of **BEAT** to change the overall speed. We can extend the usage of constants to include note frequencies as well, making our code more readable:

```
#define BEAT 50
#define A3 220
#define A4 440
PlayTone(A3, BEAT);
Wait(BEAT);
PlayTone(A4, 2*BEAT);
Wait(2*BEAT);
```

You can also patiently define a table of constants for all the notes, so you can reuse it in many different programs:

```
#define C1   33
#define Cs1 35
#define D1   37
#define Ds1 39
//...
#define C4   262
#define Cs4 277
//...
#define B8 7902
```

We coded, for example, the D# note as **Ds** (D sharp) because most languages don't allow the use of special symbols like **#** in the names of constants and variables. Don't worry about the length of this table, because constants get resolved by the compiler and don't change the length of your actual code or the space it takes up in memory.

Creating a soundtrack for your robot is a typical example of where multi-tasking proves to be really helpful. You will typically enclose your song in a separate *task*, starting and stopping it from the main task, as required by the situation.

Converting MIDI files

By using constants, your program becomes more clear, but you don't save any time in coding your melody. You still have to write the notes one by one. The good news is that some tools can do some or all the work for you. The Bricx Command Center, for instance, lets you click notes on a virtual piano keyboard on the PC screen, generating the corresponding NQC code for you. A more complete solution comes from the conversion of standard musical files.

The Musical Instruments Digital Interface (MIDI) is a complex standard that includes communication protocols between instruments and computers, hardware connections, and storage formats. A MIDI file is a song stored in a file according to the format defined by this standard.

MIDI files have achieved incredible success among professionals, amateurs, and instrument manufacturers, and are by far the most preferred way for musicians to exchange songs. For this reason, you can easily find virtually any song you're looking for already stored in a MIDI file.

But what is a MIDI file? It is simply a sequence of notes to play, their duration, their intensity, and, of course, a code that denotes the instrument to be used. Thus a MIDI file is not an audio file. It does not contain digital music like CDs, WAV files, MP3 files or other common audio formats. Rather, it contains instructions for a player (either a human being or a machine) to reproduce the song, almost a score, to be performed by actual musicians. And, as with a real score, the result rests heavily on who actually performs it. For MIDI files, this means that the output depends on the device which renders the music: with a professional MIDI cxpander you can get impressive results, while execution of the notes by a low-end PC audio card will probably be very poor. What makes MIDI files so interesting to musicians is that they are easy to read and edit (with special programs) in terms of standard musical notation.

So, the key question is: Is there a way you can render MIDI files with the RCX? Though you cannot *import* them directly to the RCX, there's a very nice utility that can convert any MIDI file into the proper code: MIDI2RCX, a free conversion utility developed by Guido Truffelli. It currently runs on Windows machines only, producing either NQC or legOS code, but Truffelli plans to implement more target languages. You can find it at Truffelli's site (see Appendix A).

Before going into the details about how to use it and what it can do for you, there's another characteristic of MIDI files you must be aware of. The notes inside a MIDI file are grouped into *channels*, and each channel is assigned to the *instrument* meant to reproduce those notes. For example, channel 1 could be assigned to an Acoustic Piano, channel 2 to a Bass, channel 3 to a Nylon String Guitar, and so on. Channel 10 is always assigned to Drums, while channel 4 is usually, but not always, assigned to the melody line, that is, the notes sung by the vocalist or played by the leading instrument. As we explained earlier, the RCX has monophonic sound capabilities and cannot reproduce more than a note at a time, so you have to choose carefully the notes it plays. Before you start converting a MIDI file into code straight away, we suggest you do some exploring using a specific software to see which channel could better render the idea of the song. There are many commercial products which are capable of manipulating MIDI files in almost every possible way, but you don't actually need all the power and complexity they provide. The Internet is crammed with freeware and shareware programs perfectly suitable for the task of identifying the best single channel to be converted into instructions for the RCX. You open your MIDI file with the editor, mute all the channels except one in turn, and decide which one to use. If you feel at ease with the MIDI editor, you can cut away some notes from the selected channel, since you probably don't need the whole song, only a chunk of it, the part that contains the refrain or main theme. If you do this through editing, you will save the modified MIDI file, of course.

NOTE

You can save a lot of work if you find a MIDI file targeted to cellular phones. These typically have sound reproduction limits very similar to those of the RCX.

Now you're ready to use MIDI2RCX. It is a console application, not a graphic interface, so you need to run it from a Command Prompt window. It requires the name of the MIDI file, and two optional parameters that specify the channel to convert (it defaults to *all*) and the target language (it defaults to *legOS*). Your typical command will be something like this:

```
c:\midi2rcx>midi2rcx letitbe.mid 4 nqc
```

where *letitbe.mid* is your original MIDI file, *4* is the converted channel, and *nqc* the target language. With this command, MIDI2RCX will produce a file named letitbe.nqc containing plain NQC code ready to be compiled, downloaded to your RCX and executed, or more likely, pasted into your own program. We strongly advise you against converting all the channels: The result will be almost unrecognizable.

Converting WAV Files

Guido Truffelli also wrote a WAV2RCX application that converts WAV files into NQC or legOS instructions. Unlike MIDI files, WAV files contain digitalized audio ready to be executed. If you are familiar with graphic file formats, you can think of MIDI files like *vector* graphics, while WAV files resemble *raster* graphics.

Sequencing MIDI files on the RCX is a challenging task. Playing a WAV file, however, is a lot more challenging. As far as we know, no one has succeeded in getting very good quality. Most likely, the RCX audio hardware has limits that aren't easy to overcome.

Truffelli's program adopts a simple strategy that leads to good results with many WAV files: It splits the source into small intervals and for each of these computes the dominant frequency using an algorithm called FFT; it then converts these frequencies into RCX program statements using the same approach as MIDI2RCX. This is not enough to make your RCX speak, but works well with simple audio patterns like the ding.wav or ringing.wav files included in the Windows system. WAV2RCX is a prized tool with which you can equip your robots with sounds in the best science fiction tradition: laser guns, jump sparks, and buzzing!

Summary

The purpose of this short journey into the sound system of the RCX was to show that, despite its strong limitations, it's still an invaluable resource. It can support you in debugging, return information in the form of sounds of different patterns or frequencies, or complete the personality of your robots.

NQC offers two commands to control the sound system: **PlaySound** to perform predefined sound patterns, and **PlayTone** to play any note of a desired pitch for the desired duration. While **PlaySound** is suitable for most user interfacing needs, **PlayTone** offers finer control and lets you create melodies.

Thanks to the work of independent developers, you can convert some of the most common digital audio formats straight into NQC instructions. Considering the hardware limitations of the RCX, MIDI files translate very well and are the ideal candidates to provide your robots with a musical soundtrack. The conversion of WAV files, on the other hand, present greater difficulties and offer poorer results; nevertheless, they can equip your robot with amazing sound effects.

More than one robot in this book relies on sounds as a relevant feature. For example, the Tic-Tac-Toe and Chess players of Chapter 20 beep to inform the user they are ready for input, and in the Flight Simulator of Chapter 24 the sound system is entrusted with an essential part of the simulation: reproducing the noise of the engine. Other robots, which can work without sound, would benefit a great deal from some sound effects—good examples of this are the animals and droids of Chapters 17 and 18. In Chapter 21, we will take a different approach, offering ideas about how to build robots capable of playing instruments themselves!

Chapter 8

Becoming Mobile

Solutions in this chapter:

- **Building a Simple Differential Drive**
- **Building a Dual Differential Drive**
- **Building a Skid-Steer Drive**
- **Building a Steering Drive**
- **Building a Tricycle Drive**
- **Building a Synchro Drive**
- **Other Configurations**

Introduction

Most robots are designed with some kind of mobility in mind. Motion makes your creatures animated and "alive," and offers a limitless number of interesting, fun, and challenging projects with which to test your creativity and skills. Most mobile robots belong to one of two categories: *wheeled* robots or *legged* robots. Though legs provide an effective way to move on rough terrains, wheels are generally much more efficient on smooth surfaces.

In this chapter, we will survey the most common wheeled mobility configurations, discussing some of their pros and cons. Please bear in mind that the chassis shown in the following examples are designed to highlight the details of gearings and connections, and for this reason, many of them need some reinforcement to be used in actual robots.

Building a Simple Differential Drive

If you have built some of the robots described in the LEGO Constructopedia, or put together the test platform outlined in Chapter 5, you're already familiar with the *differential drive* architecture. It has so many advantages, particularly in its simplicity, that it's by far the most often used configuration for LEGO mobile robots.

A differential drive is made of two parallel drive wheels on either side of the robot, powered separately, with one or more casters (pivoting wheels) which help support the weight but that have no active role (Figure 8.1). Note that it is called a differential drive because the robot motion vector results from two independent components (it's of no relation to the differential *gear*, which isn't used in this configuration).

When both the drive wheels turn in the same direction at the same speed, the robot goes straight. If the wheels rotate at the same speed but in opposite directions, the robot turns in place, pivoting around the midpoint of the line that connects the drive wheels. Table 8.1 shows the behavior of a differential drive robot according to the direction of its wheels (assuming that when it's in motion they run at the same speed).

Figure 8.1 A Simple Differential Drive

Table 8.1 Behavior of a Differential Drive Robot According to the Direction of Its Wheels

Left Wheel	Right Wheel	Robot
Stationary	Stationary	Rests stationary
Stationary	Forward	Turns counterclockwise pivoting around the left wheel
Stationary	Backward	Turns clockwise pivoting around the left wheel
Forward	Stationary	Turns clockwise pivoting around the right wheel
Forward	Forward	Goes forward
Forward	Backward	Spins clockwise in place
Backward	Stationary	Turns counterclockwise pivoting around the right wheel
Backward	Forward	Spins counterclockwise in place
Backward	Backward	Goes backward

At different combinations of speed and direction, the robot makes turns of any possible radius. This maneuverability, the capability to turn in place in particular, makes the differential drive the ideal candidate for a broad class of projects.

Add to this the fact that it is very easy to implement, and you can understand why more than 50 percent of all mobile LEGO robots belong to this category.

If *tracking* the robot position is one of your goals, again the differential drive is a good candidate, requiring very simple math. (We'll discuss this later in the book.)

There's only one real drawback to this architecture: It's not easy to get your robot to move in a perfectly straight line. Because no two motors have exactly the same efficiency, you will always have one wheel turning a bit faster than the other, thus making your robot turn slightly left or right. In some projects, this isn't a problem, particularly those programmed for continuous route correction, like following a line or finding a path through a maze. But when you want your robot to simply go straight in an open space, this problem can be really frustrating.

Keeping a Straight Path

There are many ways to maintain a straight path when using a simple differential drive. The easiest approach involves reducing the effect by choosing two motors with similar speeds. If you have more than two motors, try finding a combination with the closest matching speeds. This won't guarantee your robot actually goes straight, but it can reduce the problem to a tolerable level. We have a friend who measured the speed of his motors under a small load, and wrote the actual rpm on the bottom of each one with a permanent marker to be able to combine them with satisfactory performance.

A second simple way involves adjusting the speed via software. As described in Chapter 3, your program can control the power of each motor. You can trim the power level of the faster motor until you get an acceptable result. The problem with this approach is that when the load changes (when the robot runs on different terrains), the power levels required to maintain speed will change.

Using Sensors to Go Straight

A more sophisticated approach that has several positive side effects requires you to introduce a feedback mechanism into your system, thus controlling each wheel with sensors and adjusting their speed according to the readings. This is what most of the "real life" differential drives do. You can attach to each drive wheel an encoder that counts rotations, and then control the power level in your software to compensate for the difference in the number of turns. The LEGO rotation sensor is ideal for this task. Connect one to each wheel and measure the difference in counts, then stop or slow down the faster of the two for a while to keep the counts equal. One positive side effect is that you can use the same sensors to detect obstacles utilizing the technique described in Chapter 4. If a motor is on but the wheel doesn't rotate, you can deduce your robot is stuck against something. Another benefit is that you can use the rotation sensors to perform turns of a precise angle. Finally, they provide the basic equipment to make your robot compute its position using a technique called *odometry* which we'll discuss later in Chapter 13.

Using Gears to Go Straight

If you have only one rotation sensor, there's a little trick you can use to control the *difference* in speed between the drive wheels instead of the *actual* speed of the wheels. Recall our discussion of the differential gear in Chapter 4. You can use it to add and subtract. If you connect the drive wheels with a differential so that one wheel enters the differential with a direction that's inverted with respect to the other, the body of the differential itself should stay still when the wheels rotate at the same speed.

If there is any difference in speed, the differential gear rotates and its direction tells you which wheel is turning faster. Figure 8.2 shows a possible setup (a bit tricky, isn't it?). We strongly suggest you build this chassis even if you don't have a rotation sensor, because the mechanism is instructive and fascinating by itself. We omitted the motors and any reinforcing beams to keep the picture as clear as possible, but in your implementation you should add two motors, each one acting on its wheel like in a standard differential drive. The purpose of the geartrain on the right is to reverse the rotation direction of the axle that enters the differential gear, at the same time keeping the original gear ratio. The rotation sensor, meanwhile, connects to the body of the differential gear to detect whether it turns.

Figure 8.2 Monitoring the Difference in Right and Left Wheel Speed with a Single Rotation Sensor

A more radical solution is to lock the wheels together when you need to go straight. This system is very effective, making your robot go perfectly straight, but it requires a third motor to activate the locking system as well as some additional gearing, which makes the solution less than compact. Figure 8.3 shows an example of a locking mechanism that requires special parts: a dark gray 16t gear with clutch, a transmission driving ring, and a transmission changeover catch, which combine in a sort of clutch mechanism (Figure 8.4). That special gear has a circular hole instead of the standard cross-shaped hole, thus it rotates freely on the axle. The driving ring should then be mounted on an axle joiner. When you push the driving ring into the gear (with the help of the changeover catch), the gear becomes solid with the axle.

Figure 8.3 A Lockable Differential Drive

You can also use the setup shown in Figure 8.2, inserting a motor in place of the rotation sensor. Recall from Chapter 4 that a motor works as an electric brake, too: In its *off* state, it opposes motion, while in the *float* state it is still not powered but free to turn. In this solution, you will not power this motor, but rather operate it as an electric brake for the body of the differential. When you brake the motor in *off* state, the differential hardly turns, making your robot go straight. On the other side, with the motor in *float* state, the differential can rotate and the robot is able to turn. Table 8.2 summarizes some of the possible combinations: The rule is that when the left and right motor run with different directions, the differential gear lock motor must be in float state.

Figure 8.4 The 16t Gear with Clutch, the Transmission Driving Ring, and the Transmission Changeover Catch

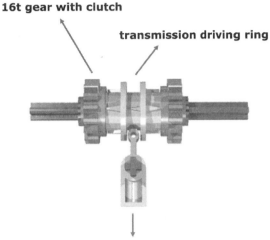

16t gear with clutch

transmission driving ring

transmission changeover catch

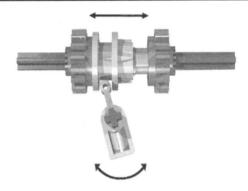

Table 8.2 How to Control a Differential Drive Robot Provided with Electric Differential Gear Lock

Left Wheel Motor	Right Wheel Motor	Differential Gear Lock Motor	Robot
Off	Off	Off	Rests stationary
Forward	Forward	Off	Goes straight forward
Forward	Reverse	Float	Spins clockwise in place

Continued

Table 8.2 Continued

Left Wheel Motor	Right Wheel Motor	Differential Gear Lock Motor	Robot
Reverse	Forward	Float	Spins counterclockwise in place
Reverse	Reverse	Off	Goes straight backward

Consider that even in float mode the motor has significant mechanical resistance, so the robot will not turn as quickly and the drive motors will be under more stress when turning.

Using Casters to Go Straight

Casters are another key factor in getting your differential drive moving and turning smoothly. Most often, though, they are not given enough consideration. The LEGO Constructopedia suggests the caster shown in Figure 8.5, but we will take the liberty of saying that it is a poorly designed caster. It uses two wheels coupled on the same axle. You already know from Chapter 2, however, that this configuration doesn't allow the wheels to turn independently. Keep the assembly gently but firmly pressed on a table, and try to rotate it in a tight turn—it doesn't turn very well, does it? In fact, unless you let one of the wheels skid, it doesn't turn at all.

Figure 8.5 The Coupled Caster from Constructopedia

The casters shown in Figure 8.6 get much better results. The one on the left uses a single wheel, thus avoiding the problem entirely. The one on the right, which is more solid, uses two free wheels that allow the caster to turn in place without friction or slippage problems. The difference is in the wheel hubs. In the assembly on the left, the axle turns with the wheel, while the one on the right has the wheels spinning on the axle.

Figure 8.6 Casters Designed to Avoid Skidding

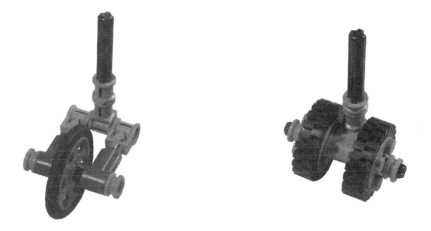

The choice of using one or more casters depends on what task the robot is designed for. A single caster is enough for most applications, but two casters at the front and rear of the robot are a better option when stability is important.

In some cases, as with a simple robot of limited weight that has a smooth surface on which to navigate, you can substitute the caster with *inverted round tiles* or other parts that provide limited friction when contacting the floor (Figure 8.7).

Figure 8.7 Inverted Round Tiles Can Replace Casters

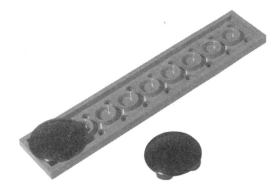

Building a Dual Differential Drive

A *dual differential drive* is an improvement on the simple differential drive. It is designed to mechanically solve the problem of following a straight path, and uses only two motors (see Figure 8.8). Its gearing setup is a bit complex, and relies again on the differential gear—two of them to be precise (see Chapter 9 about getting supplementary parts).

Figure 8.8 A Dual Differential Drive

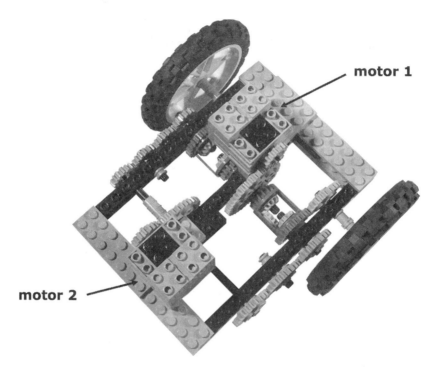

The dual differential drive inverts the common use of the differential gear. Normally, the wheels are connected to the *axles* coming out of the differential gear, while in this case, the wheels are connected to the *body* of two differential gears. In Chapter 4, we explained that a differential gear can be used to mechanically add or subtract two independent motions; to do this, use the axles coming out of the differential gear as *input*, and the body of the differential gear will move according to the result of their algebraic sum (a sum that takes direction into account).

In this setup, both motors provide one input to the two differential gears. The trick is that one of the motors rotates the input axles of the two differentials in the same direction, while the other is geared to rotate the other input axles in opposite directions. To operate a dual differential drive, you will normally use just one of the motors, keeping the other braked.

In Figure 8.9, you see the same assembly as in Figure 8.8, but without motors. When motor 1 rotates the 40t gear A, and motor 2 keeps B braked, motion gets transmitted along the dotted line path in the picture, the two differentials rotate in sync and the robot goes straight. On the other hand, keeping motor 1 off and consequently A braked, and operating motor 2 to rotate B will make the motion transfer along the solid line and the differentials rotate at the same speed, but in opposite directions. The result is that the robot spins perfectly in place.

Figure 8.9 The Dual Differential Drive Dissected

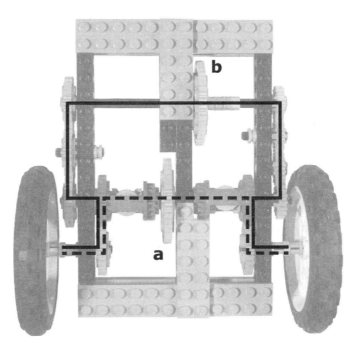

Thus, you would normally use a single motor at a time, one for going straight, the other for turning. Nothing bad happens if you power both motors—depending on their direction. One of the differentials will receive two opposing inputs, nullifying them and remaining stationary, while the other adds two inputs, doubling the resulting speed, in which case the robot pivots around the stationary wheel, exactly like a simple differential drive does when one of its wheels moves and the other rests.

A very nice feature of the dual differential drive is that with a single rotation sensor you can precisely monitor any kind of movement of your robot. Couple the sensor to one of the wheels (it doesn't matter which one). When the robot goes straight, you can use the sensor to measure the traveled distance, and when the robot turns in place, the sensor measures the change in heading.

Of course, remember we said earlier that there are no free lunches in mechanics. In other words, this ingenious configuration has its drawbacks. The first, obviously, is its complexity. We deliberately built our example flat on a plane to keep all the connections easy to understand; however, you can build more compact versions by stacking some of the gearing (it will still require all those gear wheels, maybe just a couple less). The complex gearing leads to the second side effect: our nemesis *friction*. To make matters worse in this case, you have just a single motor to fight it!

Building a Skid-Steer Drive

A *skid-steer drive* is a variation of the differential drive. It's normally used with tracked vehicles, but sometimes with 4- or 6-wheel platforms as well. For tracked vehicles, this drive is the only possible driving scheme. Good examples of skid-steer drives in real life are excavators, tanks, and a few high-end lawnmowers.

Figure 8.10 shows a simple tracked skid-steer drive. Each track is powered by its independent motor, that mounts an 8t gear and meshes a 24t gear connected to the track wheel. The front track wheels need not be powered.

Figure 8.10 A Tracked Skid-Steer Drive

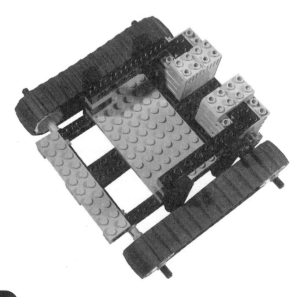

A wheeled skid-steer drive requires a trickier setup. You must transmit the power to all the wheels, otherwise your platform won't turn smoothly, or might not even turn at all. The model shown in Figure 8.11 uses a row of five meshed 24t gears for each side, all of them receiving power from two motors like in the tracked version. Every wheel axle mounts its gear, and they are interleaved with idler gears that serve the purpose of transferring motion from one wheel to the other. If you do have enough 24t gears, you can mix them with 24t crown gears, which are exactly the same size. The balloon tires in the picture come from supplementary sets.

Figure 8.11 A Wheeled Skid-Steer Drive

Tracked robots are easy to build and fun to see in action, thus placing them among the favorites of many builders. Just as with differential drives, when the tracks go the same direction, the robot goes forward; differences in their speeds or directions make the robot turn; in-place steering is possible, too. Skid-steer drives also share with differential drives the same difficulties in getting them to move in a straight line.

Here is where the similarities end, and some peculiarities of skid-steer emerge:

- Tracks have a better grip than wheels do on rough floors and terrains, but this is *not* true on smooth surfaces.

- Tracks introduce more friction which uses up some of the power supplied by the motors.

- The unavoidable skidding intrinsic in the nature of these vehicles makes them absolutely unsuitable for applications where you need to determine the position by utilizing the motion of the robot.

Building a Steering Drive

A *steering drive* is the standard configuration used in cars and most other vehicles that features two front steering wheels and two fixed rear wheels. Thankfully, it's suitable for robots too. You can drive either the rear or the front wheels, or all four of them, but the first is by far the easiest solution to implement with LEGO parts, so this is what we'll cover here. Though less versatile than differential drives, and impossible to steer in place or in very tight turns, this configuration has many advantages: It's very easy to drive straight, and very stable on rough terrain.

When building a steering drive robot from the basic MINDSTORMS equipment, you have only one motor to power the drive wheels, because you need the other to steer the front wheels. Thus your steering drive robot will have about half the power of a differential drive one, which can benefit from both motors during straight motion.

In Figures 8.12 and 8.13 you see two simple steering platforms. Apart from implementation details, these two models share the same construction principles. For instance, the rear wheels are connected to the driving motor through a differential gear. As explained in Chapter 2, you cannot avoid the differential if you want your vehicle to turn. A second motor steers the front wheels, providing your robot with a way to change direction. Notice that we used a belt to drive the steering mechanism, taking advantage of its implicit torque-limiting transmission to avoid any damage to the mechanism or the motor if the motor remains on after the steering mechanism has reached one of its limits. You would probably add a sensor to detect the steering position, allowing your robot to control its direction. A single touch sensor is the bare minimum needed—make it close when the steering is centered, so you can use timing to steer the wheels and utilize the sensor to center them back after the turn (Chapter 14 contains an example of this technique).

Figure 8.12 A MINDSTORMS-only Steering Drive

Figure 8.13 Another Steering Drive

Using Ackerman Steering for Smooth Turns

True-life steering vehicles implement a more sophisticated scheme called *Ackerman steering* (from the name of the person who first studied it). In our simple design, the steering wheels turn at the same angle, but this is not entirely correct—during turns, the inner wheel goes along a tighter bend than the outer one. During large radius turns, the difference is small and its effect negligible. In tight turns, however, the effect becomes quite noticeable, causing one of the steering wheels to skid. Ackerman's steering system is designed to compensate for the different turning angle of the inside wheel, thus eliminating any skidding. The theory says that the vehicle turns smoothly when the "lines" extended from every wheel axle meet and revolve around one common point (Figure 8.14).

Figure 8.14 Ackerman Steering Scheme: The Inner Wheel Turns More than the Outer One

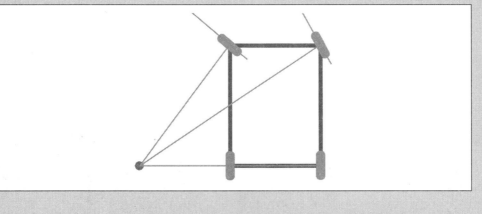

Building an Ackerman scheme with LEGO is definitely possible. Chapter 14 incorporates the prototype of a front-wheel drive that features the Ackerman correction.

Both models employ a *rack and pinion* steering mechanism where an 8t gear (the *pinion*) meshes with a special plate with teeth, a sort of "unrolled gear" (the *rack*). The difference between the chassis in Figure 8.12 and the one in Figure 8.13 is that we built the latter using extra parts that make our life easier: three 1 x 10

TECHNIC plates, two steering arms, and two tiles. These components are designed to be combined together, creating a very simple steering mechanism used in many LEGO TECHNIC car and truck models. In the model presented in Figure 8.12, built only from MINDSTORMS parts, we had to use a 2 x 8 plate, instead of the 1 x 10 ones, and replace the steering arms with a home-made version. The whole front section of the vehicle has been built with the beams oriented studs-front, to provide the necessary support for the wheels and the steering mechanism, but mostly to provide a smooth surface (the side of the beam) which the rack can slide over (you will find more information about this setup in Chapter 14).

When you build the steering assembly, you can move the wheel behind its pivoting axle for self-centering steering (an advisable property in many situations). In version a in Figure 8.15, you see a wheel mounted just below the pivoting axle, which does not effect the steering. If you mount the wheel behind its steering column, friction causes the dynamic forward motion of the car to push the wheels toward the rear, resulting in a self-centering action. Look at the design of a shopping cart, and you will see that the actual wheel contact area is behind the pivoting axis. The more you move the wheel behind the pivoting axis, like in versions b and c, the more self-centering you get. Don't ever mount the wheel in front of the pivoting axle, like in version d. This will make your steering unstable. In fact, the wheel will tend to go toward the rear, causing your car to turn spontaneously.

Figure 8.15 Moving the Wheel from the Pivoting Axle

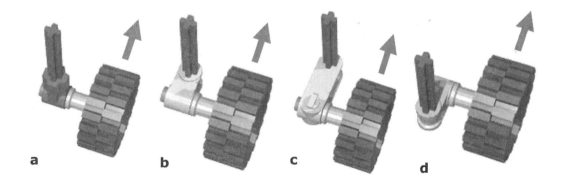

a   b   c   d

We encourage you to experiment with these concepts, building a simple chassis and exploring the properties of the various assemblies shown in Figure 8.15.

The steering drive is a suitable configuration for rough terrains, since it's very stable on its four wheels. You can improve the grip of the wheels on the ground by using some kind of suspension. It's very important that none of the drive wheels permanently lose contact with the ground, otherwise the differential would find the path of least resistance and transfer all the power to that wheel, resulting in the wheel spinning and your robot becoming immobilized.

A *limited slip differential* can help reduce this problem (see Figure 8.16) by connecting the wheel axles to a common supplementary axle through pulleys and belts. The belts tend to keep the driven axles rotating at the same speed, but during turns they slip a bit on their pulleys, allowing the wheel to adjust their speeds. Should a wheel lose contact with the ground, the belts will still be able to transfer a good portion of power to the other wheel.

Figure 8.16 A Limited Slip Differential

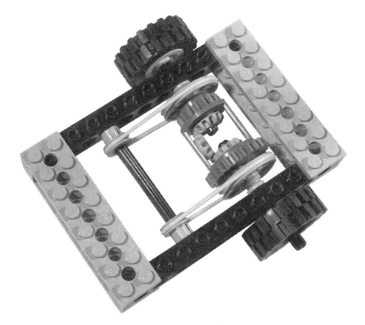

Building a Tricycle Drive

A *tricycle drive* configuration involves a front wheel that drives and steers and is matched with two passive independent rear wheels which provide stability (Figure 8.17). The peculiarity of this configuration lies in the fact that the front wheel is both powered and steering, giving the robot a high grade of mobility.

Figure 8.17 A Tricycle Drive

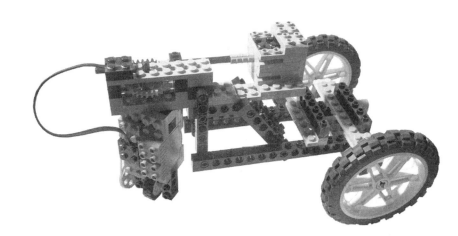

You might think that driving the rear wheels instead of the front one would give you the same results, but this is true only for a limited range of steering angles. In fact, like in a steering drive, when narrowing the turn radius, you ultimately reach a point where the rear wheels can no longer convert power into motion. The maximum turning angle that a steering vehicle can reach is when the inner wheel is stationary and the outer one draws a circle around that point. A front-wheel driven tricycle, on the other hand, can manage any steering angle, even when the wheel is perpendicular to the direction of motion of the rear wheels.

Ideally, the driven wheel can rotate 360° to point in any possible direction. This means you should build a system with no constraints on a full turn (an example of this architecture is the mechanism used to drive bumper cars at amusement parks). Our example in Figure 8.14 is capable of rotating the steering a full 360°, but cannot make more than a single 360° rotation due to the wire that connects the motor to the RCX.

In practical applications, a 180° turn is enough to allow the robot any possible movement, because any angle in the range of 180° to 360° is equivalent to an angle in the range of 0° to 180° with the motion reversed. In other words, 210° with the motor in forward motion corresponds to 30° (210 − 180 = 30) with the motor in reverse. As with the steering drive, you will probably use a sensor to detect the position of the steering.

Building a Synchro Drive

A *synchro drive* uses three or more wheels, all of them driven and steering. They all turn together in sync, always remaining parallel, thus the robot changes its direction of motion without changing its orientation.

Synchro drives are quite challenging to build with LEGO parts. Until a few years ago, there was general agreement that it should have been possible, yet nobody had succeeded in the undertaking. Now the barrier has been broken, and if you navigate the Internet, you can find many well-designed LEGO synchro drives.

To make a full 360° synchro drive and avoid any limitations in its turning ability, the key point is to transfer motion along the pivoting axle of each wheel. The simplest approach requires a special part called the *turntable*, a large round rotating platform usually employed in LEGO models to support revolving cranes or excavators (Figure 8.18).

Figure 8.18 The LEGO Turntable

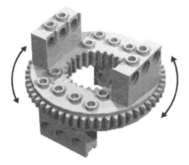

You can attach the wheel to one side, and drive it with an axle that passes through the hole in the center of the turntable. In Figure 8.19, you can see an example of this technique. Notice that the turntable is upside down, because the wheel must be connected to the part of the turntable that gets rotated by the external gear. Because of this, the robot will result in an entirely, or at least partly, studs-down design!

We want our synchro drive robot to be able to change direction in place without moving. To this aim, the two assemblies in Figure 8.19 and 8.20 are similar, but not interchangeable. With the driving axle blocked, the lower part of the turntable should turn smoothly in place—in Figure 8.19 it does, but in Figure 8.20 it doesn't. This happens because the wheel in Figure 8.20 is not centered below the pivoting axle, and so when it changes its direction it has to travel some distance. The gearing in Figure 8.19 makes the wheel rotate in the proper direction, the one that complies with the turn, while the gearing in Figure 8.20 makes the wheel oppose the turn. We realize this is a subtle difference, and we invite you once again to learn by experience, building the two versions by yourself and verifying how they work.

Figure 8.19 A Possible Wheel Assembly for a Synchro Drive

Figure 8.20 Incorrect Version of the Wheel Assembly

To build a complete synchro drive, you need at least three of these turntables. Then you have to connect them so that one motor can drive all the axles at the same time, while another can turn all the wheels in sync.

In Figure 8.21 you see the bottom view of a four-wheeled synchro drive. Notice that we linked the turntables with 8t gears so they all turn together. Powering any one of those 8t is enough to make the robot change direction.

Figure 8.21 A Complete Synchro Drive (Bottom View)

Figure 8.22 shows the top view of the same platform: the large 40t gear (a) drives the wheels through four pairs of bevel gears, while the other 40t (b) is in charge of turning the wheels. To complete this synchro, you have to add two motors to power a and b, possibly using an 8t gear to get a ratio which is capable of reducing the friction introduced by all that gearing.

Synchro drives are quite amazing to see in action, and yours will be no exception. But if you expect it to navigate the room detecting obstacles, your challenge isn't quite over yet: You still have to manage bumpers. In a synchro drive, the concept of "front" and "rear" has no meaning, since the robot can travel using any of its sides as a front. Consequently, you have to place bumpers all around it. As you learned in Chapter 4, if your robot has four sides, you are not compelled to use four sensors for four ports (which your RCX doesn't have). You can connect four touch sensors to the same port, using an OR configuration by which any sensor that gets closed puts the RCX into an "on" state. Or you could simply use a single omni-directional sensor like the one shown in Figure 8.23; the touch sensor is normally closed, but opens whenever the upper axle departs from its default position (kept by the rubber bands). Surround your robot with a ring of tubes or axles, connect the ring to the omni-directional sensor, and that's it!

Figure 8.22 A Complete Synchro Drive (Top View)

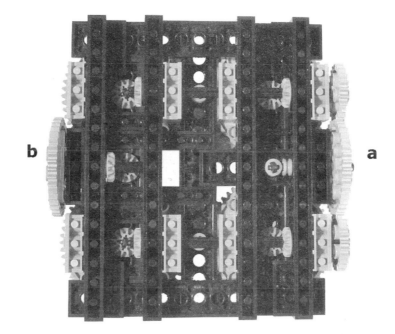

Figure 8.23 An Omni-Directional Touch Sensor

Other Configurations

Our roundup doesn't cover all the possible mobile configurations. There are other more sophisticated or specialized types:

- **Multi-Degree-of-Freedom (MDOF) vehicles** MDOF vehicles have three or more wheels, or groups of wheels, both independently turned and driven. Imagine a synchro drive where each wheel can change its speed and direction with no connection to the others: such a robot would be able to behave like a differential drive, a steering drive, or a synchro drive just by controlling its configuration from the software. Though interesting to study and very versatile in their use, they are also extremely difficult to build and control. In fact, not all of their possible configurations result in a coordinated motion!

- **Articulated Drive** This is very similar to the steering drive, but instead of steering the wheels, it steers a whole section of the vehicle. The front wheels always remain parallel to the front part of the chassis, and the same applies to the rear wheels in regards to the rear portion of the chassis. Nevertheless, the two sections connect through an articulation point that lets them pivot in the middle. This configuration is common in wheeled excavators and other construction equipment.

- **Pivot Drive** Keith Kotay defines a *pivot drive* as a configuration made of a chassis with non-pivoting wheels with a platform in the middle that can be lowered or raised. When the platform is up, the robot moves perfectly straight on its wheels. When it requires turning, it stops and lowers the platform until the wheels don't touch the ground anymore. At this point it rotates the platform to change its heading, then raises the platform again and resumes a straight motion.

- **Tri-Star Wheel Drive** The Tri-Star configuration has been designed for high-mobility, all-terrain vehicles. Each "wheel" is actually an equilateral triangle with wheels in each vertex; the vehicle features three of them for a total of twelve wheels. The wheels turn, and the triangles can also turn like larger wheels. During normal motion, two wheels of each triangle touch the ground, but when a wheel sticks against an obstacle, a complex gearing system transfers motion to the triangular structure, which turns and places its upper wheel past the obstacle. As complicated to build as it is interesting!

- **Killough Platform** Developed by Francois Pin and Stephen Killough, the official name of this mechanical configuration is Omnidirectional Holonomic Platform (OHP). *Holonomy* is the capability of a system to move toward any given direction while simultaneously rotating. While conventional wheeled vehicles aren't holonomic at all, this platform allows for unprecedented mobility. Seen from the top, a Killough drive shows three wheels placed at the vertices of an equilateral triangle. Each "wheel" is a sort of sphere made of actual wheels combined together and used in a quite unconventional way: on their side!

We hope we've made you curious about these configurations, and invite you to find out more about them using the reference material provided in Appendix A. All of them can be built from LEGO parts, and give you further challenges for when the standard configurations shown in this chapter have become old hat.

Summary

This chapter has been quite dense, but we hope we've been able to help you in choosing a drive configuration. When building a mobile robot, different architectures are relevant to its resulting shape, and most importantly, to its performance.

The differential drive is simple and versatile, but can't go straight. The steering drive, meanwhile, goes straight but cannot turn in place. The dual differential drive can do both, but it's more cumbersome and complex to build. Robotics is like cooking: there are many recipes for the same dish, but to be successful you still must know the ingredients well and use them in the right proportions. Of course, don't forget to add the most important ingredient of all: your creativity.

Chapter 9

Expanding Your Options with Kits and Creative Solutions

Solutions in this chapter:

- **Acquiring More Parts**
- **Creating Custom Components**
- **Creative Solutions When More RCX Ports Are Needed**

Introduction

If, by now, you are caught up in robotics, you may feel a bit constrained by the limitations of the MINDSTORMS kit. You want more. What do you perceive most limiting: the number and range of parts, or the fact your RCX has only three input and three output ports? Maybe you would like to use new kinds of sensors, or servo-motors. And why not try out some pneumatic devices?

If the MINDSTORMS was your first LEGO set, you will be pleased to see that there are many additional parts to boost and support your creativity. If MINDSTORMS is an addition to your large collection of LEGO TECHNIC sets, you already know what parts the line includes and probably already have them—but there is a also wealth of compatible non-LEGO custom parts and kits you may never have dreamed of: infrared and ultrasonic proximity detectors, compasses, sound frequency decoders, magnetic switches, and voice recognition units, just to mention a few. In this chapter, we will explore some options for expanding your designs and plans, surveying the most important additions, providing you with information about where and how you can get them, and describing also the most significant non-LEGO custom devices.

Extra parts are not the only way to expand your project ideas. Some mechanical tricks can also help you in getting the most from the limited number of output ports offered by the RCX. You will learn how a single motor can power two or more mechanisms, and how you can apply this trick to some of the mobile configurations we described in Chapter 8.

Acquiring More Parts

Describing all the components that make up the LEGO world would be a tremendously difficult task. The vast LEGO system includes tens of thousands of different parts, belonging to different themes, but all are easily integrated with each other. That's the beauty of LEGO: You can always find a new use for something that might have been built with a completely different purpose in mind. Whether it be towns, trains, or pirates, any or all of the LEGO themes might add something useful to your set of equipment. Of course, when it comes to building *robotics*, the natural choice is the LEGO TECHNIC line.

Created in 1977 to introduce older children to the world of mechanics and motors, the TECHNIC line developed into a complete system that includes many specialized parts. You are already familiar with the almost 140 varieties found in the MINDSTORMS kit, organized into some of the classes previously mentioned—beams, plates, axles, liftarms, gears, and so on.

Introducing Some Specialized Components

Many other TECHNIC parts come in a broader variety than shown in the MINDSTORMS kit. Liftarms, for example, are increasingly prevalent in recent TECHNIC releases (Figure 9.1). There's an evident trend in this direction, and in fact, some of the newest sets don't include traditional beams or plates at all. They're instead composed only of liftarms, axles, and connectors.

Figure 9.1 Liftarms

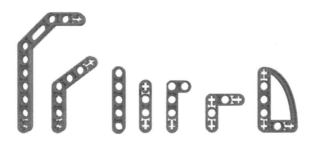

Liftarms have many possible uses. We showed in Chapter 5 that they can profitably replace beams to brace the layers of a structure, especially in those cases when you need a vertical lock that remains within the height of the horizontal beams because you have other plates or beams above or below them. Other common transformers include levers and arms, since their cross-shaped holes provide an ideal attachment point for axles when you need to operate them through some kind of mechanism.

In previous chapters, we covered some representatives of the new class of gears not included in the MINDSTORMS kit, like the 20t bevel gear, the 20t and 12t double-bevel gears, and the 16t gear with clutch (Figure 9.2). There isn't currently a service pack specific to gears only, and to increase your inventory, you have to buy TECHNIC models or MINDSTORMS expansion sets, which include many other parts.

Figure 9.2 Gears Not Included in the MINDSTORMS Kit

Special components called *gearboxes* help you in assembling compact and solid gearings (Figure 9.3). Version a combines a worm gear with a 24t, thus performing a 1:24 reduction. We explained in Chapter 2 that worm gears can *transfer* motion but not *receive* motion—in other words, they can turn a gear but cannot be turned by it. We also explained that this feature is of great help when you want a mechanism to rest in its current position when you've switched the power off—for example, in an arm aimed to lift weights. This gearbox is ideal for such high-torque applications, because it encloses the 24t and the worm gear into a solid body case where the gears cannot fall apart.

Version b, the newest of the three, integrates well with standard beams and provides a convenient way to change the direction of motion or to split power onto two or three axles using 12t bevel gears.

Version c comes from older TECHNIC sets and corresponds to b in the way of functionalities. It's a bit harder to integrate with other parts, but has the advantage of allowing vertical mounts for the gears.

Figure 9.3 Gearboxes

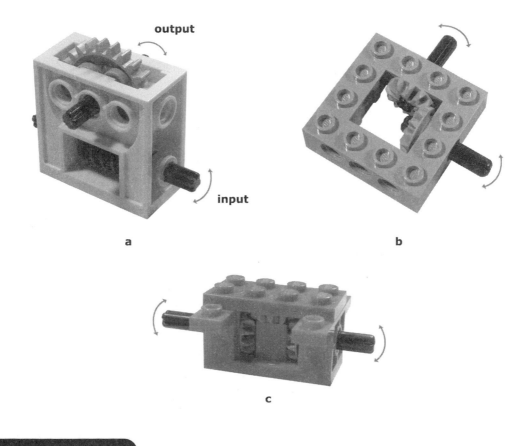

a

b

c

Chain links, another component not included in the MINDSTORMS kit, come in two types. The first is meant for use in transmissions, as explained in Chapter 2 (*chain links*), while the second was designed to make up tracks of arbitrary length (*track links,* see Figure 2.19 in Chapter 2). They're a nice feature, although it's a pity the tracks' links don't get a better grip on most surfaces and that they come apart rather easily.

In Chapter 8, you were introduced to the turntable (see Figure 8.15 in Chapter 8), a very useful part for building rotating subassemblies, as well as some TECHNIC plates and connectors specially suited for building steering assemblies in association with the rack gear. Figure 9.4 shows how to combine three 1 x 10 TECHNIC plates, a rack gear, two 8t gears, two steering arms and some axles and connectors into a fully functional rack and pinion steering assembly. Two of the plates brace the beams of the chassis, enclosing the steering arms at their ends. The third plate connects the free ends of the steering arms and lets them pivot while remaining parallel with each other; a rack on the plate meshes with the pinion connected to the steering motor (not visible). Notice that the pinion is made up of two 8t gears; one wouldn't be enough, because while steering, the plate would move toward the stationary plates and it would lose contact with a single 8t gear pinion. Mount the wheels on the steering arms using *axles with a stud.*

Figure 9.4 TECHNIC Plates and Connectors for Building a Steering Assembly

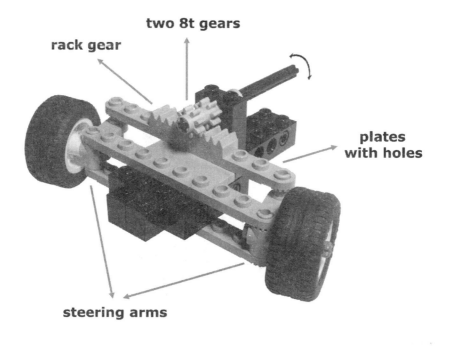

two 8t gears

rack gear

plates with holes

steering arms

The movable plate is supported by two 1 x 4 *tiles*, which are like plates with no studs and which provide the ideal smooth surface other parts can slide over (Figure 9.5). Racks and tiles are a very good combination for producing linear motion, and we confess we've never understood why LEGO didn't put any tiles in the MINDSTORMS kit.

Figure 9.5 Tiles Are Like Plates with No Studs

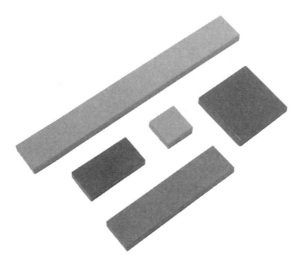

Some of the largest and most famous TECHNIC sets reproduce cars. They are a useful source for, among other things, *shock absorbers* (Figure 9.6), and large wheels (Figure 9.7). The use of shock absorbers is not limited to their traditional function, that is, keeping the wheels of a vehicle in touch with the ground on uneven or rough terrains; you can profitably employ them as springs in many kinds of mechanisms, including bumpers.

Figure 9.6 A Shock Absorber

Figure 9.7 The Wheel from the 8448 Super Street Sensation

The *flex system* (Figure 9.8) allows the transference of linear motion from one point to another distant one, exactly like the wire ropes that control the accelerator and clutch in motorcycles, or the brakes on a bicycle. You probably won't need them very often, but they allow you to operate a part through a distant motor. They also prove extremely useful in making compact and lightweight mechanisms—for example, you can open/close a robotic hand at the end of a long arm that extends from a motor placed in the main body of the robot.

Figure 9.8 The Flex System

LEGO TECHNICS also features a line of *pneumatic devices*: small and large cylinders, small and large pumps, pipes, and valves (Figure 9.9). They offer so many possibilities when it comes to robotics that we decided to dedicate an entire chapter to them (Chapter 10).

Figure 9.9 Components of the Pneumatic System

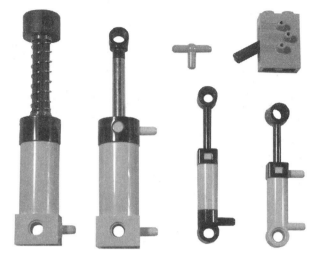

Designing & Planning…

Choosing Colors

Most TECHNIC parts come in a large assortment of colors, which include the traditional LEGO colors (white, red, blue, yellow, green, black, and gray) and some more recent ones (tan, dark gray, light blue, light-green, lime, purple, orange, and brown). If you care about colors as much as we do, this is great news; a conscious use of colors can improve the appearance of robots. However, there's much more that colors can do for you: they can help in making the structure and the mechanisms of your robot more evident. In our favorite scheme, we use two colors for the body of the robot: one for plates and another for beams, liftarms, and all the other static parts. This makes the layered structure very easy to read. Then we use one or two additional colors for mobile parts to highlight their function in the robot. For example, the fact that you employ a beam as a connecting rod between two parts of a mechanism is more apparent if its color stands out against the prevalent colors of the robot.

In large and complex robots, you can use colors to identify its subsystems: one color for the mobile platform, another for the grabbing arm, a third for the rotating head, and so on for each relevant unit.

Continued

Colors help also in keeping the wiring neat when necessary. A dual RCX robot, for example, can use up to 12 input and output connections, and some of these wires are probably not so easy to trace inside the structure of the robot. Place pairs of small plates on the connectors at both ends of a wire, using a different color for each wire, and you'll have no problem understanding which port is connected to the motor and which is connected to the sensor.

Buying Additional Parts

Now that you've seen all these parts, you might wonder where you can get them. This is a very good question which, unfortunately, has no easy answer. There are general accessory sets, themed sets, expansion sets, and service packs, as well as general catalogues. Each may offer more or less than you need at one time, and price may also be a factor.

The MINDSTORMS line has many sets, but in our opinion some of them are priced a bit too high for their actual value. The 3801 Ultimate Accessory Set is a good choice, including a rotation sensor, a touch sensor, a light brick, a remote control, and other parts.

The 9732 Extreme Creatures Set contains few interesting parts for its price, but remember the Fiber Optic System unit, as explained in Chapter 4, can be used as a rotation sensor, too. The 9730 RoboSports Set is a bit more interesting, as it contains an extra motor. The most notable parts contained in the 9736 Exploration Mars Set are two gearboxes, six balloon tires, two very long cables (3m) and a bunch of beams, plates, gears, and connectors. In our opinion, these three sets are good purchases only if you find them at a reduced price.

The 9735 Robotics Discovery Set contains a unit called *Scout* that's a sort of younger brother of the RCX. Scout incorporates a light sensor, and features two output ports for motors and two input ports for sensors (passive types only: touch and temperature). It has a large display and offers some limited programmability from its console, without the need for a PC, thus offering an easy start to robotics. Despite this nice characteristic, we feel it's a bit too limited.

The two Star Wars MINDSTORMS sets, the 9748 Droid Developer Kit and the 9754 Dark Side Developer Kit contain an even more limited unit, MicroScout, that incorporates a motor and a light sensor, but has no ports. It has seven predefined programs, and can be interfaced to the Scout with an optical link to act as its third motor. Through the Scout you can also download a tiny program to the MicroScout. In our opinion, MicroScout is definitely too simple to be used for robotics, so we again suggest you buy these sets for their parts, and only if you find them for sale at a discounted price. If you really want another programmable brick, we strongly recommend a second MINDSTORMS kit, which with its RCX, two motors, three sensors, and more than 700 additional parts, in our opinion remains your best option.

LEGO also released a video camera system called 9731 Vision Command. The camera connects to your PC, and a dedicated LEGO software can send IR commands to your RCX unit, through the tower, according to what happens inside the observed area. Don't dream of recognizing shapes or performing other sophisticated artificial vision tasks, since Vision Command allows only very basic reactions to changes in some predefined areas of the screen. You will discover also that the cable that links the camera to the PC is a constraint to your robot mobility. Despite these limitations, however, Vision Command opens up a world of possibilities.

MINDSTORMS expansion sets are an option, and TECHNIC sets another. Sad to say, but the current TECHNIC line does not include many expansion sets with suitable parts for robotics. Old TECHNIC sets had more beams and plates then current ones do, which, as we explained, tend to rely more and more on studless liftarms, which are useful but somewhat complicated to use. If you are so lucky as to find some discontinued TECHNIC sets, you have a good chance of it better suiting your needs. Being bound to the current production, large sets are a better purchase than small ones, having a higher ratio between functional and decorative parts. We prefer not to suggest any specific model here, as each fan has his or her own preferences; also, every year LEGO releases new sets and discontinues others.

With all that said, it is perfectly understandable that you may simply wish to buy only the specific parts you need. LEGO offers a mail service, called Shop-At-Home, from whose catalog you can order both sets and *elements packs* or *service packs*. Recently LEGO started an online service called LEGO Direct, through which you can order from your computer, pay with your credit cards, and get the parts or sets shipped to your door. LEGO Direct has been greeted with great enthusiasm by LEGO fans who see it as the promising beginning of a new era, one where everybody can order only the specific parts they need from a complete catalog. Currently, LEGO Direct offers the current line of sets and a limited choice of service packs, but the range is increasing and we all hope that it ends in a thorough and practical worldwide service.

Another useful resource is the DACTA service. DACTA is the branch of LEGO devoted to educational products, whose catalog includes a wide range of sets and supplementary kits. Though packed with a different assortment, the DACTA boxes contain the same parts used in commercial LEGO products. In all countries, the sale of the DACTA line is entrusted to companies specialized in selling educational items to institutions, though they normally sell to the public, too (for example, PITSCO in the USA and Spectrum Educational in Canada). Though not exactly cheap, the DACTA catalog includes many parts no longer available in sets or service packs, like the turntable or the track links, and many other parts that remain hard to find in large quantities, like the 40t gear and the rotation sensor.

Last but not least, LEGO fans from all over the world have formed a sort of community that has its own selling services. Some fan-run Web sites offer an impressive array of new and used parts and sets, in either mint or used condition, and most of the sellers accept credit cards and ship internationally. See Appendix A for some links to these commercial and private Internet LEGO shops.

Creating Custom Components

In the following sections, you will see that some of the proposed enhancements involve parts not supplied by the LEGO company. This applies in particular to electronics like motors and sensors.

We understand that your attitude toward non-LEGO parts could range from enthusiasm to hostility. You might see the benefit in making your own temperature sensor (spending only $2 instead of the $30 that the original costs), or you might be keen on the opportunity of giving your robot a voice recognition device. On the other hand, you might feel that using non-LEGO parts is a violation of the rules of the game, or you may be so fond of LEGO that you wish not to contaminate it with foreign components.

We can not, and will not, recommend one viewpoint over the other—the choice must be yours. We are personally open to some nonoriginal devices, provided that they "look like" LEGO parts. These can be cased into LEGO bricks, use standard LEGO wires and connectors, and quite closely resemble the originals. However, the use of aluminum plates, brass nuts, and bolts through LEGO beams, axles or plates cut to match a specific size, and visible chips and resistors are all unacceptable options to us. This is, again, our own choice, however.

Limiting your choices to LEGO parts has a certain appeal. It's like a common paradigm inside which you challenge yourself and other people to reach higher and higher goals. Most of the time, we build pure LEGO robots, using other devices only when we have something special in mind that we feel can really benefit from that particular hardware. Staying with original LEGO is particularly important when approaching contests and public challenges. It's a simple way to regulate what's admitted and what's not, and to be sure, too, that all competitors are pulling from identical resources.

On the other side, if you're open to experimenting with non-LEGO devices, your horizons become much broader. In this section, we'll provide some examples of what can be done with them, our assumption being that you continue to use LEGO parts to build your robots, and the RCX to run them; thus, we'll discuss the use of non-LEGO sensors and motors only.

The LEGO company doesn't release much information about the internals of its electronic devices, so most of the technical details currently available to the public are based on the work of the pioneer hackers who analyzed and dissected the sensors and motors. Michael Gasperi is the person who made the strongest single contribution to this process, his Web site and book being reference points for any work in the field. Some of these custom devices are really easy to make if you can solder, or have a friend who can. In this chapter, we will show you some of *what* can be done; refer to Appendix A to find resources that teach you *how* to make this stuff, or tell you where to buy it.

Building Custom Sensors

Michael Gasperi 's site explains how to build some simple custom sensors. The simplest of all is probably the passive light sensor built with a cadmium sulfide (CdS) photo-resistor and nothing more (Figure 9.10). This sensor is much better than the original LEGO light sensor in measuring ambient light, though it's a bit slow in acknowledging variations. With two CdS cells and some electronics, you can build a differential light sensor, which tells you in a single value if there's any difference in the amount of light received by the two units; this is very useful in pinpointing light sources.

Figure 9.10 Single and Double CdS Light Sensors

Recycling junk is an option when trying to save money. Figure 9.11 shows a touch sensor made with a switch from a computer mouse. Pulling apart a broken mouse, you will discover that there are some micro-switches connected to its push-buttons. Unsolder them from their circuit plate, solder their terminal to an electric plate, then add some parts to the case in the switch.

Figure 9.11 A Mouse Switch Recycled into a Touch Sensor

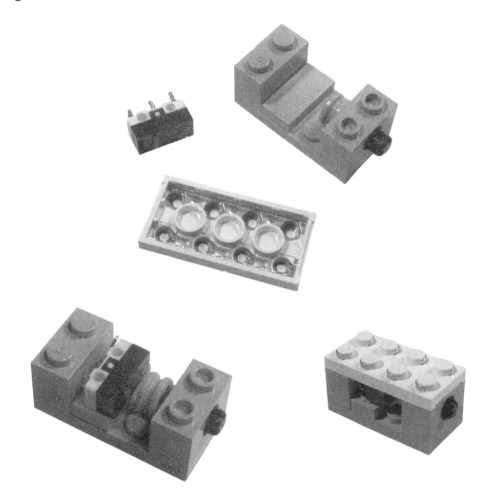

There are many people who describe in their Web sites how to make custom sensors, providing schematics and detailed instructions. Some of them also sell construction kits or finished sensors. Pete Sevcik is a good example of this latter category; his sensors are very well engineered and professionally cased into LEGO bricks. Figure 9.12 shows three of his *infrared proximity detectors* (IRPD). An IRPD is a sensor based on the IR light proximity measurement system we explained in Chapter 4, with the advantage being that you are not required to do anything in your code, just read the sensor value. IRPD sensors have an incredible range of applications. They are perfect for obstacle detection, of course, but you can use them also to make your robot follow your hand movements, to trigger the grabbing feature of a robotic hand, to find soda cans or locate your opponent during competitions. As we explained in Chapter 4, the proximity detection technique cannot measure distances, but it can tell you if an object is coming closer or entering its field of detection. The rightmost sensor in Figure 9.12 is a *dual IRPD*, able to detect an obstacle within a wider angle and tell you if it's front, left, or right with a single reading.

Figure 9.12 Different Kinds of Infrared Proximity Sensors

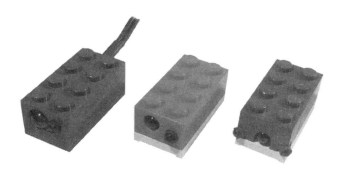

Sevick also produces a *pitch sensor*, a sophisticated sound sensor that returns a value proportional to the frequency of the incoming sound. You can thus control your robot by simply whistling or playing a flute or recorder like a modern Pied Piper. The robotic pianist of Chapter 21 represents a possible application for this sensor: It can learn a simple melody just by listening to it.

John Barnes is another person who has shown incredible creativity and competence in building custom sensors. Barnes made one of the first LEGO compatible ultrasonic sensors (Figure 9.13), a device able to measure distances evaluating the delay between the emission of a sound and its returning echo. Like a sonar, the sensor emits an ultrasonic signal (not audible), reads its echo, and returns a value that represents the distance of the closest object. The fields of application of these sensors are similar to what's described for the IRPD sensors, with the further advantage that ultrasonic sensors return an absolute distance value. This means that your robot can improve its navigation abilities, because it can not only avoid obstacles but also find the best route evaluating the distances of the surrounding objects.

Figure 9.13 An Ultrasonic Distance Sensor

Barnes has assembled many other amazing devices, including a *compass* with a resolution of 3.75° (Figure 9.14) and a *pyroelectric sensor* able to detect the presence of humans or animals by measuring the changes in ambient IR radiation (Figure 9.15). The compass sensor just looks like a pile of bricks, because there isn't any device emerging from its body, but the inside contains a small electronic compass and a circuit to convert its output into values that the RCX can interpret. Connect the compass sensor to an input port of the RCX configured for a light sensor, and it will return values in the range of 0 to 95, where 0 is North, 24 is East, 48 is South, and 72 is West. Having the RCX know which way it's pointing in order to keep going straight and having it make known angle turns makes a big difference in solving navigation problems!

Figure 9.14 A Compass Sensor

Figure 9.15 A Pyroelectric Sensor

The casing around the pyroelectric sensor has a small hole that lets its internal "eye" receive the infrared light any warm body produces. It requires some time to adapt to the ambient radiation, but afterward it can detect any change in intensity. These features make it unsuitable for mobile robots, but it's very useful in those projects where a robot must start doing something when it detects a human presence.

Probably the most astonishing of Barnes' devices is his Voice Recognition unit (Figure 9.16). After a short teaching session, you will be able to give simple one- or two-word commands to your robot like "go," "stop," "left," "take" and see your robot perform the required task. It's rather large and heavy, because it contains its own set of batteries, and, consequently, is not very easy to place in a compact robot. However, it gives reality to the dreams of robots harbored by every sci-fi fan: the ability to respond to vocal commands!

Figure 9.16 John Barnes' Voice Recognition Unit

No-contact switches are very useful tools, too. These are switches that close without the need of physical contacts with the casing of the sensor. We integrated Michael Gasperi's General Purpose Analog Interface with a *Hall-effect detector* to build a sensor capable of detecting magnetic fields (Figure 9.17). A Hall-effect detector is a small integrated circuit which returns different signals depending on whether it is in the presence of a strong magnetic field or not. Gluing a small permanent magnet on a LEGO peg, you can easily mount it on any mobile part of the robot. When the magnet comes very close to the sensor, the latter detects it.

Figure 9.17 A Hall-Effect Sensor

Chris Phillips followed a simpler and more effective approach to get the same result using a cheap and easy-to-mount *Reed switch*. A Reed switch is a small bulb containing two thin metal plates very close to each other. When you put the bulb close to the source of a strong magnetic field, the metal plates touch and complete the circuit. Small permanent magnets are the ideal parts to trigger this sensor, with the same procedure we described for the Hall-effect sensor. You can also use the LEGO magnets designed to couple train cars. Detecting trains is actually what Phillips devised his sensor for, but it is suitable for many other applications: It can replace touch sensors in almost all applications, and even emulate rotation sensors if you mount the permanent magnet on a wheel that makes it pass periodically in front of the sensor.

Figure 9.18 shows a Reed bulb mounted in series with a 100K resistor over a LEGO *electric plate*, which provides an easy way to interface custom sensors to the standard LEGO 9v wiring system. The final sensor will be cased in a hollowed brick to make it look like a standard LEGO component.

Figure 9.18 A Reed Switch Sensor before Final Assembly

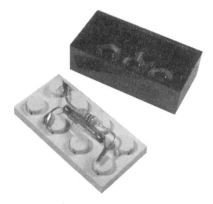

Solving Port Limitations

Some of the electronic devices that have appeared in the LEGO robotics community are meant to solve the endless dilemma of the limited input and output port number. The common approach involves *multiplexing*, a technique through which signals from different sources are combined into a single signal. Michael Gasperi explains how to build a very simple multiplexer that can host up to three touch sensors and return a value that the RCX decodes into their respective states (Figure 9.19). This device takes advantage of the fact that the RCX can read raw values instead of simple on/off states, and returns a unique number for any possible combination of three sensors.

Figure 9.19 A Three Touch Sensor Multiplexer

Nitin Patil designed a more complex multiplexer suitable for connecting a single input port to three active sensors, like the original light and rotation sensors, or any other custom active sensor like IRPDs, sound, and so on. Active sensors use the entire raw value range, thus this device cannot combine their signals into a single number like the three touch sensor multiplexer does. Actually Patil's device connects a single sensor at a time to the port, and requires the RCX to send a short impulse to select the desired sensor (Figure 9.20).

Figure 9.20 A Three Active Sensor Multiplexer

Pete Sevcik's Limit Switch, though not a multiplexer, allows you to save some ports by combining two touch sensors and a motor on a single output port (Figure 9.21). Until a switch closes, the motor is under normal control from the RCX. When a touch sensor gets pressed, the inner circuit prevents the motor from turning into a specific direction, thus automatically limiting the motion of a mechanical device. If your robot has a rotating head, this limit switch can make it stop at its left and right bounds using just a single port.

Figure 9.21 Pete Sevcik's Limit Switch

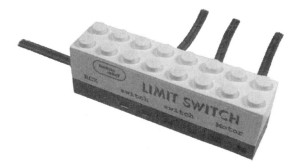

Output port multiplexing, though technically possible, doesn't get the same attention as input port multiplexing, thus there are few schematics and little documentation on this topic. The focus seems most on using different kinds of motors, *servo motors* in particular. Servos are typically used in radio-controlled models to steer vehicles, move ailerons, and handle other movable components. They are cheap and have high torque, thus they are ideal for some applications. Unfortunately, they expect power in a specific waveform that the RCX cannot supply. Ralph Hempel solved the puzzle creating a simple electronic interface that performs the appropriate conversion, thus revealing the power of servo motors to LEGO robotics hobbyists.

NOTE

The number of electronic expansion devices is vast, and still growing. If you are curious about these devices, we once again invite you to visit some of the Web links we provide in Appendix A.

Creative Solutions When More RCX Ports Are Needed

When you start gaining experience with LEGO robotics, and wish to build something more complex than your early robots, you will quickly find yourself facing the heavy constraints imposed by the limited number of ports the RCX has. Are three motors and three sensors too few for you? If you feel a bit frustrated, remember that you're in good company. Thousands of other MINDSTORMS fans feel the same!

In Chapter 4, we provided some tips on connecting more sensors to a single input port. We are going to describe here some tricks that, using only LEGO components, allow you to somewhat expand your motor outputs.

Start by observing that in some applications you don't need a motor turning in both directions, just one motor in one direction. Your robot can take advantage of this fact by driving two different gearings with a single motor. Figure 9.22 shows how you can achieve this using a differential gear: Its output axles mount two 24t gears that can rotate each one only in a single direction. The two 1 x 4 beams work like *ratchets*. They let the gear turn in one direction but block its teeth in the other. If you connect the motor to the body of the differential, it will drive either the right or the left axle depending on its direction.

Another setup, shown in Figure 9.23, is based on the fact that the worm gear is free to slide along the axle.

Figure 9.22 Splitting a Rotary Motion on Two Axles

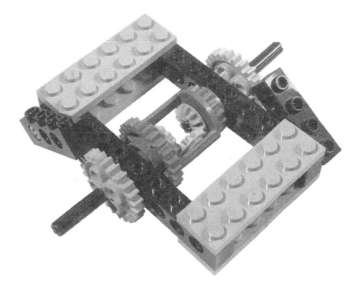

Figure 9.23 The Crawling Worm Gear

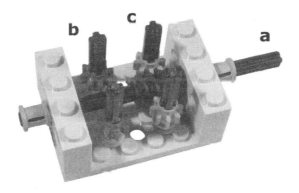

   Provided that there is some friction in the output axles B and C, when axle A turns clockwise, the worm crawls left until it engages the B 8t gears and gets stopped by the beam. Turning A counterclockwise, the worm crawls right, disengaging the B gears and engaging the C pair. Thus, with a single input axle you get two pairs of outputs, each pair having one axle turning clockwise and the other counterclockwise. We invite you once again to build and test this simple assembly. It's almost unbelievable to see!

To put theory into practice, let's see how you apply these principles to the mobility configurations of Chapter 8. The differential drive is a good starting point. Can you drive two wheels with a single motor? Yes, you can—using the differential gear to split its power onto two separate outputs. Then, copying the design of Figure 9.22, add a ratchet beam that acts on one of the wheels (Figure 9.24). The motor drives both wheels through the differential when going forward, but one of them gets blocked during reverse motion, making the robot pivot around it. Simple, but limited. It's not guaranteed to go straight, and cannot spin in place. Nevertheless, it allows you to make a mobile platform that uses only one port of your RCX!

Figure 9.24 A Single Motor Differential Drive

The dual differential drive shown in Chapter 8 is a good starting point for a more sophisticated solution. Its design uses one motor to drive straight and the other to change direction. You should replace these motors with a mechanism similar to that of Figure 9.22, making the output axles of its differential gear (the third of the robot!) take the place of the motor shafts. Now apply a motor to the last differential gear: In one direction it will make the robot go forward, in the other it will make the robot spin in place. It works, though we realize that the resulting gearing probably isn't the simplest thing we've ever seen!

Even in the synchro drive (Figure 9.25) you can get full motion control with a single motor. Relying on the fact that the synchro drive has the freedom to turn on its wheels at any angle, you can keep them turned in the same direction until they reach the desired position. Again, apply the scheme of Figure 9.22 and make one output axle of the differential gear operate the steering mechanism, while the other provides drive motion.

When the motor turns one way, the wheels change their orientation, and when the motor turns in the other direction, the wheels move the robot forward. Backward motion is not required, because the wheels can point to any heading and the motion reversal is performed by a 180° change in their direction. With a platform like this, you have complete control over navigation, and you still have two free output ports to drive other devices.

Figure 9.25 A Single Motor Synchro Drive

Single motor tricycle drives are possible, too, requiring a gearing very similar to that of our single motor synchro drive. Make just one of those steering-driven wheels, add two rear free wheels, and you're done.

This trick of splitting the turning directions over two separate axles obviously won't cover all your needs for extra ports. In many cases, you must control both directions of your gearings, but you probably don't need all motors running at the same time. In a robotic arm with three independent movements, for example, you use three motors, but using just one at a time doesn't affect its global functionality. The idea is to use one motor to make a second motor switch among several possible outputs. This approach will always require two motors, and engages two output ports, but can give you a virtually unlimited number of independent bi-directional outputs, only one of them running at any time. A possible implementation of such a device is shown in Figure 9.26. The motor at the bottom drives five 16t gears all linked together. On the other side of the assembly there are five 8t output gears not connected to the previous 16t. A second motor at the top slides a switching rack that, through a 16t on one side and a 24t on the other, connects the input gears to one of the five possible outputs. We used a touch sensor to control the position of the switching rack: five black pegs close the switch in turn when the gears are in one of the five matching positions. Due to its large size, this setup is probably more suitable for static robots than for mobile ones.

Figure 9.26 Switching a Motor among Five Output Axles

The previous example requires two output ports and one input port. In this case, as well as some others, we can save an input port by implementing a sort of *stepper motor*. A stepper motor is a motor that, under a given impulse, turns precisely at a known angle, just a single step of a turn. Stepper motors are widespread devices. You can find them in any computer printer or plotter, and in digital machine tools. LEGO doesn't make a stepper motor, nor does the RCX have dedicated instructions for them, but Robert Munafo found a pure-LEGO solution. Our version is a slight variation of Robert's original setup (see Figure 9.27). A rubber band keeps the output axle down in its default position. You have to power the motor for a short time, enough to make the axle get past the resistance of the rubber band and make a bit more than half a turn. Now put the motor in float mode, wait another short interval, and let the rubber band complete the turn of the axle. For any impulse made of a run time and a float time, the output shaft makes exactly one turn.

Figure 9.27 A Stepper Motor

The beauty of the system is that timing is not at all critical. The on time can be any interval that makes the axle rotate more than half a turn but less than one and a half, while the float time can be any interval equal to or greater than the time needed for the rubber band to return to its default position.

Summary

In this chapter, we have been discussing extra parts, expansion sets, custom sensors, and tricks for using the same motor for more than one task:

- Extra parts come from either sets or service packs. Unfortunately, it's not always easy to buy just the parts you need, because sometimes they don't come in a service pack and you have to buy a set that contains many other elements you don't need. The LEGO Direct Internet shop is growing quickly, and it promises to become a very thorough and practical service. DACTA supplier and fan-run online shops fill the gap in the offer of spare parts, giving you countless opportunities to improve your equipment set.

- Custom sensors are a new frontier, and reveal a whole new world of possibilities. Would you like your robot to measure the distances of the objects around it? It's possible. Would you like it to recognize vocal commands? Again, it can be done. Proximity detectors, sound sensors, magnetic switches, electronic compasses, input multiplexers… the Internet is crowded with Web sites that teach you how to build your own MIND-STORMS-compatible custom sensors, or that sell them ready to use.

- Mechanical tricks enable you to use the same motor to power multiple mechanisms. Through the use of a differential gear and a couple of ratchet beams, you can split the output of a motor between two output axles. This principle extends to the point of driving a complete platform with a single motor.

There's a common denominator for these apparently unconnected topics— we want to push the limits farther. What this means (and can mean) depends on you, on what your rules are in regards to using non-LEGO parts, on how much you can spend on expansion sets, and how imaginative you are in finding new solutions to problems. Don't give up without a fight! Reverse the problem, or start again from scratch, or let the problem rest for a while before you attack it again. Look around you for inspiration, and talk to friends. Most of the greatest MINDSTORMS robots ever seen came from ideas that seemed impossible at first glance.

Chapter 10

Getting Pumped: Pneumatics

Solutions in this chapter:

- **Recalling Some Basic Science**
- **Pumps and Cylinders**
- **Controlling the Airflow**
- **Building Air Compressors**
- **Building a Pneumatic Engine**

Introduction

In Chapter 9, we mentioned that *pneumatics* might be a nice addition to your robotic equipment. Now, we'll explore the topic in more detail. Pneumatics is the discipline that describes gas flows and how to use its properties to transmit energy or convert the same into force and motion. Most pneumatic applications use that gaseous mixture most widely available—air—and the LEGO world is no exception.

Pneumatics is a great tool for robotics, and is especially useful when your mechanisms need linear motion or an elastic behavior. It's also very useful as a way to store energy for subsequent uses. We will briefly cover the basic concepts of pneumatics, then put those theories into practice, explaining how LEGO pneumatic components work and what you can do with them, along the way showing you how to stop and start airflow in order to produce motion in your robot. By the end of the chapter, you should hopefully be up to speed on many pneumatic components, including: valves, pumps, cylinders, compressors, and pneumatic engines.

Recalling Some Basic Science

To understand pneumatics, you have to recall the properties of gases. The most important property is that they have neither specific shape nor volume, because they expand and fill all available space within a container. This means that the quantity of gas inside a tank does not solely depend on the tank's volume. The greater the quantity of gas in a given volume, the higher its *pressure*.

NOTE

The science that describes the properties of gases is called *thermodynamics*. Its *Ideal Gas Law* relates four quantities: volume, pressure, temperature, and mass (expressed in moles). In our simplified discussion, we will deliberately ignore temperature, since, in our situation, it shall essentially remain constant throughout.

We all have the opportunities to experiment with pneumatics using everyday objects. The tires of a bicycle are a good example: Their inner volume is constant, but you can increase their pressure by pumping air in. The more air inside, the greater the pressure, and the more it opposes external forces—in other words, the tires become harder.

This example leads to a second important property of compressed gases: Their pushing outward on the walls of their containers illustrates their *elasticity*. Elasticity is the property of an object that allows it to return to its original shape after deformation. The greater the elasticity, the more precisely it returns to its original configuration. In the example of the bicycle tire, if you push your finger against it, you can temporarily create a dimple in the surface, but as soon as you remove the finger, the tire resumes its shape—the greater the pressure inside, the higher the resistance to deformation.

The fact that gases are so easy to compress is what makes pneumatics different from *hydraulics* (the science of liquid flow). Essentially, liquids are uncompressible.

When you compress a gas into a tank, increasing its pressure, you are storing energy. Pressure can be interpreted also as a *density of energy*, that is, the quantity of energy per unit volume. This leads to a very interesting application of pneumatics: You can use tanks to accumulate energy, which can then be later released when needed. You pump gas in to increase the pressure in the tank, storing energy, and draw gas out to use that energy, converting it into motion.

A flow of air or gas in general is produced by a difference in pressure: The air flows from the container with the higher pressure into the one with the lower pressure, until the two equalize. (In this context, we've given the term *container* the widest possible meaning. It can be a tank, a pipe, or the inner chambers of a pump or cylinder.)

Pumps and Cylinders

LEGO introduced the first pneumatic devices in the TECHNIC line during the mid-eighties, then a few years later modified the system to make it more complete and efficient. After a long tradition of impressive pneumatic TECHNIC sets, including crane trucks, excavators, and bulldozers, they discontinued the production of air-powered models. Fortunately, LEGO pneumatic devices have been recently reissued in a specific service pack (#5218) available through Shop-At-Home or at the LEGO Internet shop.

The basic components of the LEGO pneumatic systems are *pumps* and *cylinders* (see Figure 10.1). The function of a pump is to convert mechanical work into air pressure. They come in two kinds, the large variety, designed to be used by hand, and its smaller cousin, suitable for operation with a motor. Cylinders, on the other hand, convert air pressure back into mechanical work, and come in two different sizes as well.

Figure 10.1 Pumps and Cylinders

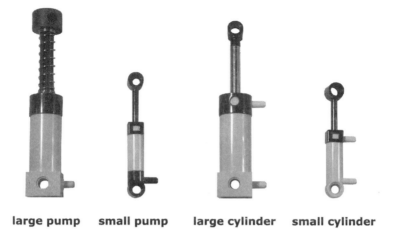

large pump   small pump   large cylinder   small cylinder

Figure 10.2 shows a cutaway of the large pump in action. When you press its piston down, you reduce the volume of the interior section, thus increasing the pressure and forcing air to exit the output port until the inner pressure equals that outside. When you release the piston, the spring pushes the piston up again; a valve closes the output port so as not to let the compressed air come back inside the pump, while another valve lets new air come in around the piston-rod. The small pump follows the same working scheme exactly, with the difference being that it doesn't contain a spring and its piston needs to be pulled after having been pushed. It's designed to be operated through an electric motor.

Cylinders are slightly different from pumps. Their top is airtight and doesn't let air escape from around the piston-rod. The piston divides the cylinder into both a lower and upper chamber, each one provided with a port. The basic property of a pneumatic cylinder is that its piston tends to move according to the difference in pressure between the chambers, expanding the volume of the one with higher pressure and reducing the other until the two pressures equalize, or until the piston comes to the end of its stroke. When you connect the lower port to a pump using a tube, and supply compressed air into the lower chamber, its pressure pushes the piston up. Doing this, the volume of the chamber increases, and this lowers the pressure until it's equal to that of the upper chamber. During the operation, the port of the upper chamber has been left open, so its air can freely escape, reaching equilibrium with the outside air pressure. Similarly, when you connect the upper port to the pump, and supply compressed air, the piston moves down (Figure 10.3).

Figure 10.2 Cutaway of the Large Pump in Action

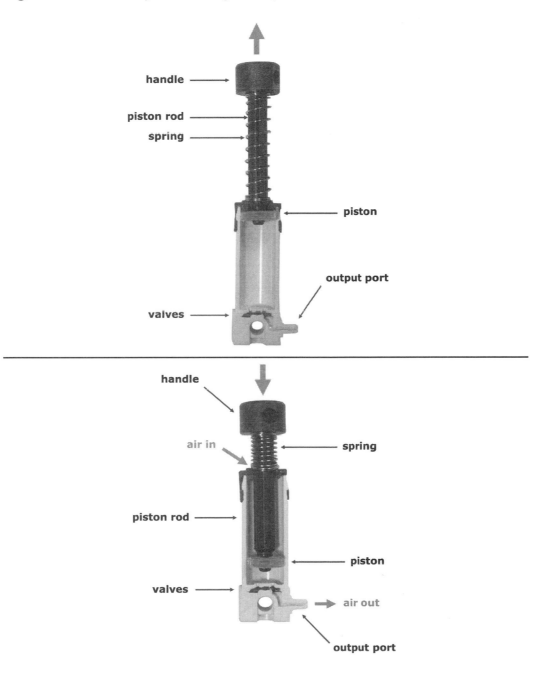

Figure 10.3 Cutaway of the Large Cylinder in Action

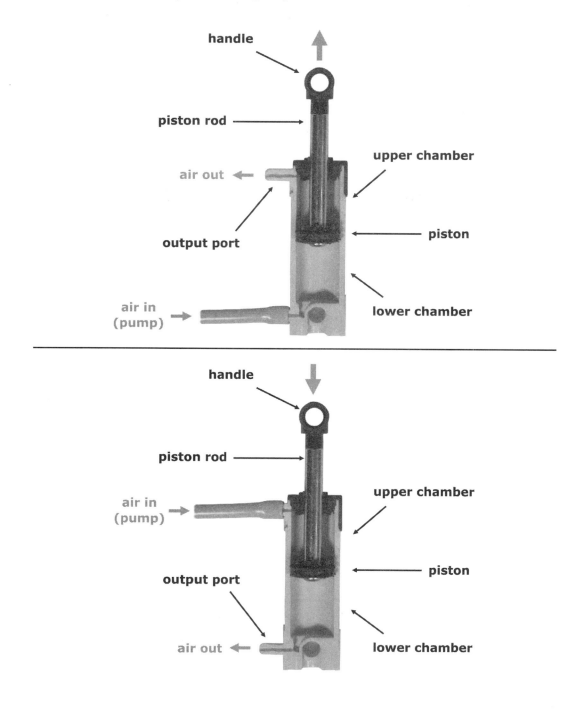

Surely you don't want to move the tube from one port or the other to operate the cylinder. It may work, but it's not very practical. The LEGO *valve* has been designed precisely for this task: It can direct the airflow coming from a pump to any one of the two ports of a cylinder, while at the same time let the pressure from the other chamber of the cylinder discharge into the atmosphere (see Figure 10.4). The valve also has a central (neutral) position, which traps the air in the system so the cylinder can neither move up nor down.

Figure 10.4 The Basic Pneumatic Connection

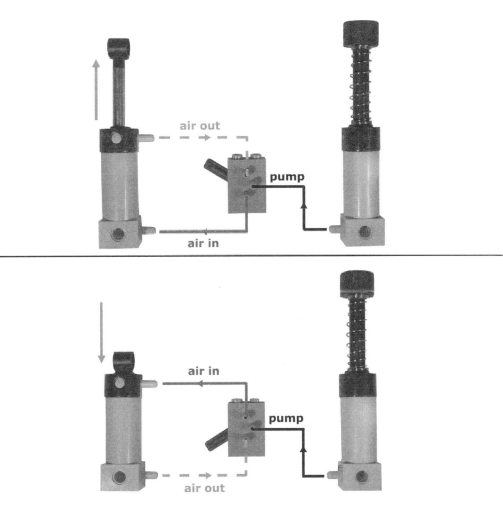

The LEGO tubing system is completed by a *T-junction* and a *tank* (see Figure 10.5). T-junctions allow you to branch tubes, typically to bring air from the source to more than a single valve. The tank is very useful for storing a small quantity of compressed air to be used later. We explained that increasing pressure is like storing energy, thus the air tank can be effectively considered an accumulator: charge it with compressed air and release it through the valve when necessary to convert that energy into mechanical work.

Figure 10.5 A T-Junction and a Tank

Pneumatic cylinders provide high-power linear motion, and thus are the ideal choice for a broad range of applications: articulated arms or legs, hands, pliers, cranes, and much more. In describing the basic concepts about pneumatics, we told you that compressed gases tend to make their containers react elastically to external forces. You can test this property with LEGO cylinders, too: connect a cylinder to a pump and operate the pump until the piston of the cylinder extends in full. Now, press the rod of the cylinder. You can push it down, but as soon as you stop applying force, the rod comes back up again. This property is quite desirable in many situations.

Let's suppose you're going to build a robotic hand. If you try to use an electric motor to open and close the hand, you must somehow know when to stop it. To do this, you can use some kind of sensor as a feedback control system that tells your RCX the object has been grabbed and the motor can be stopped. However, a pneumatic cylinder, in most cases, needs no feedback. The air pressure closes the hand until it encounters enough resistance to stop it. This approach works in a wide variety of objects. (If your robot is designed to hold eggs, be sure it exerts a very gentle pressure!) Figure 10.6 shows a simple pneumatic hand grabbing different kinds of objects. You see that we used a scissor-like setup that gives our hand a rather large range in regards to the size of the things it can handle.

The previous example gives you an idea of what pneumatics can be used for. Likely, you're already imagining other interesting applications. Unfortunately, the LEGO pneumatic system was not designed to be electrically controlled, so to effectively use it in your robotic projects you need an interface that allows your

RCX to open and close valves. And, unless you plan to run behind your robot pumping like crazy, you probably would like to provide it with an automatic compressor.

Figure 10.6 A Simple Pneumatic Hand

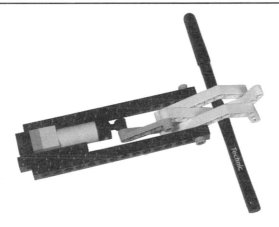

Controlling the Airflow

Almost every LEGO robotics fan would like LEGO to release an electric valve to control pneumatic cylinders, but until it does, you have to get by with mechanical solutions.

What you need in this case is a kind of indirect control similar to the one we showed in Chapter 3 when talking about the polarity switch. Figure 10.7 shows one of many possible solutions: The motor turns the clutch gear through a crown gear; on the same axle of the clutch gear there's a liftarm that operates the valve. We used the clutch gear as usual to make the timing less critical and avoid any motor problems should it stay on a bit longer than required. You can use a standard 24t gear as well. This might even be necessary if you find a valve stiffer than average. They're not all the same, and some are really hard to operate.

Figure 10.7 An Electric Valve

The downside of our electric valve is that it's able to switch between the two outermost positions, but centering the valve in its neutral position is almost impossible. This is not a big problem, because in most applications you can leave the cylinders connected to the air supply. However, if you really need the central position, you can use a touch sensor that controls when the valve is centered, and utilize a slightly slower gearing, like that shown in Figure 10.8.

Figure 10.8 The Electric Valve with a Sensor

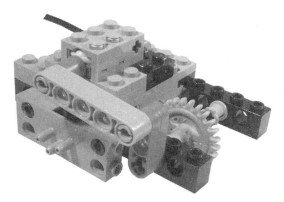

This electric valve is not very compact, but there's not much more you can do considering the size of the LEGO motor. Just the same, it works rather well, and you may feel satisfied with it. But could you make something better? Try applying some of the tricks you learned in previous chapters. For example, you know you can control more than one valve with a single motor. You have seen that, using a differential, it's possible to separate the two turning directions of a motor on two different axles. Now you only need to connect each axle to a valve so the valve cycles between its positions using only one turning direction. This is done using a liftarm as a connecting rod, like in old steam locomotives (see Figure 10.9).

Figure 10.9 A Cycling Valve

Figure 10.10 shows a prototype of a complete double electric valve, which combines two setups like those of Figure 10.9 with a motor, a differential gear and some additional gearing. We had to use a worm gear to drive the body of the differential because this mechanism requires a lot of torque to be operated. The differential splits the power onto two 40t gears, each one featuring a ratchet beam that lets it rotate only in a specific direction. Thus, when the motor turns clockwise, one valve moves, while if it turns counterclockwise, the other does, each one cycling between all positions.

Figure 10.10 A Single Motor Dual Electric Valve

There's one last problem to solve: How do you know which state each valve is in? And how do you stop the motor precisely when a valve reaches the desired position? Timing it is not an option. As we've said before, it's very difficult to control mechanisms through timing only. You can use sensors, of course, but where should you put them?

Two touch sensors per valve would add up to four sensors; that's a bit too much for your RCX, and anyway you are looking for an option that conserves ports. If you use just two touch sensors, putting them so they close when the valves are at an extreme, you can use timing to go to the other position. In this case, you would avoid timing errors, because you have a sensor that gives you absolute positioning. Can you connect the two sensors to the same port? No, because you wouldn't be able to tell which valve closed the sensor.

What if you placed the sensor so it's closed when the valves are centered? You are not going to use that position to control the pneumatics, but it would still be useful as a reference point for positioning. In order to change the position of the valve, your code has to drive the valve until the sensor closes, and from then on keep the motor running for a small interval to reach the limit point. As in the previous case, you partly rely on timing, but without cumulative errors. The advantage of this configuration is that you can connect both sensors to a single port, because they only close as they pass through a reference point.

So, we finally figured out how to use one input and one output port instead of two output ports. It's one alternative, and not a big advantage—you can still make it better. If you could just count motor rotations without using a sensor but you can! Do you remember the stepper motor from Chapter 9? Using that configuration, you can avoid using any sensors, thus fully operating two valves with a single motor! The resulting, rather complicated setup is shown in Figure 10.11.

To be honest, we've never used such a thing in any model. It's more of an academic issue, used here to make you understand there are always many ways to solve a problem, and many different paths by which to reach your goals.

Building Air Compressors

Now that you have discovered a way to operate pneumatic cylinders from your RCX, the next step is to provide them with a good supply of compressed air. Some applications require only a small quantity of air for each motion, in which case you have the option to preload a tank by pumping it manually before you run the robot. A good example is a robot that blows out a candle. All it has to do is find the candle in the room, then release its air supply to blow it out. You can extend the range using more tanks (In Chapter 27, we'll describe a robot with seven air tanks.), but for most practical applications you will need something more substantial: an unlimited source of compressed air.

Figure 10.11 A Stepper Motor Dual Electric Valve

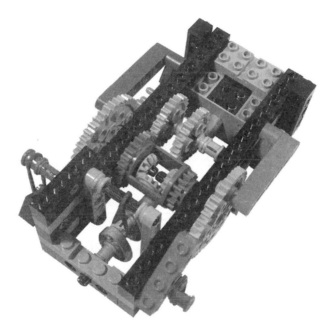

This goal is easily achieved by building an electric compressor, like the one shown in Figure 10.12. The small LEGO pump is connected to a pair of pulleys mounted on the shaft of a motor. There are many possible setups, but it's very important you design yours to take advantage of the entire stroke of the pump, because this will make it more efficient. In fact, if your compressor, for example, uses half of the stroke of the pump, it will release only half the maximum quantity of air it could potentially release. In our example, we adjusted the distance using a 1 x 2 two-hole beam, but there are many other possibilities.

The whole LEGO robotics community is grateful to C.S. Soh, who carefully tested many different compressors, some using two or even four pumps, others using the large hand pump with the spring removed. Using a pressure sensor connected to the RCX, he tested all the common designs and published the results on his site, which, by the way, contains a huge amount of information about LEGO pneumatics in general.

Figure 10.12 A Simple Compressor

According to Soh's results, the most efficient design is a slightly modified version of Ralph Hempel's compressor (see Figure 10.13). It uses two small pumps and belongs to the category of *double acting compressors*, meaning that one of the pumps takes air in while the other is pumping, thus providing a continuous flow.

Figure 10.13 Ralph Hempel's Double Acting Compressor

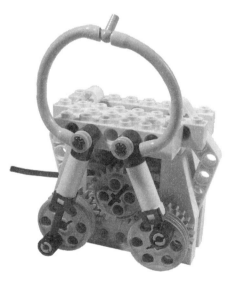

In Figure 10.14, you see another double acting compressor with a different design but with an efficiency comparable to Hempel's.

Figure 10.14 Another Double Acting Compressor

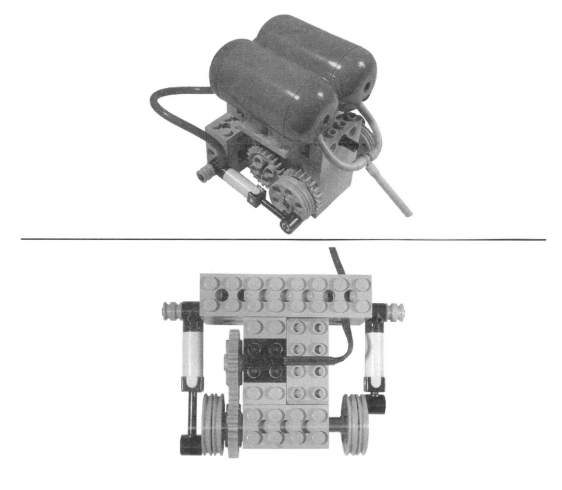

The nice thing about compressors is that they don't need to be wired to one of the precious output ports of your RCX: A battery box is enough to run them. But you might wonder when you should stop your compressor, and how. The simplest option is not to stop it. Instead, you can place a torque-limiting component in the gearing, like a pulley or a clutch gear, so that when the pressure reached a given level, the gearing idles. A much more elegant solution again comes from Ralph Hempel and is shown in Figure 10.15.

This clever pressure switch is built around a LEGO polarity switch, a small cylinder, two rubber bands, and some structure beams and plates. The bottom cylinder inlet connects to the air supply circuit of your pneumatic system, and as the pressure increases, the cylinder starts overcoming the resistance of the rubber bands. The movable part of the cylinder connects to two liftarms that operate the polarity switch, one side of which is wired to the battery box, while the other is hooked to the compressor. The polarity switch has three positions: forward, off, and reverse. In this application, you use the first two of them. When the cylinder is retracted, the polarity switch connects the battery box to the compressor. Just before the cylinder reaches its maximum extension, the polarity switch turns off, thus stopping the motor. By adjusting the number and strength of the rubber bands, you can set your pressure switch for the maximum desired pressure, complementing your compressor in a totally automatic system.

Figure 10.15 A Pressure Switch

Building a Pneumatic Engine

We mentioned before that you can make cylinders control other cylinders. This is accomplished by making a cylinder operate the valve that controls a second cylinder. This is not useful in itself, but you can make a cylinder do something *and* move a valve. One very interesting case is one in which you connect two cylinders in a loop where each one controls the other, resulting in an unstable system that continuously, and automatically, changes its state (Figure 10.16). Provided that you have a supply of compressed air, you can take advantage of this feature to make your robot perform an action.

Figure 10.17 shows a diagram of this pneumatic circuit. Cylinder 1 operates valve 1, which controls cylinder 2, which operates valve 2, which controls cylinder 1!

Probably the first robot based on this system to appear publicly on the Internet was Bert van Dam's pneumatic insect. Our slightly modified replica is shown in Figure 10.18.

Figure 10.16 An Unstable Pneumatic System

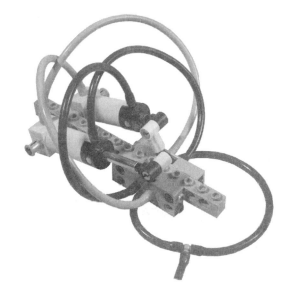

Figure 10.17 Diagram of the Cyclic Pneumatic System

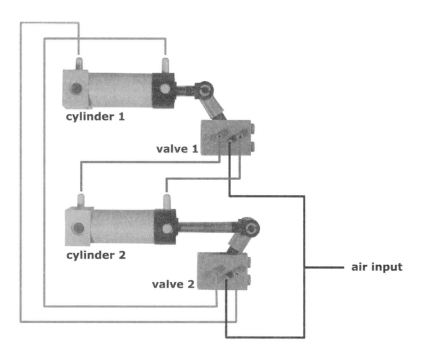

Figure 10.18 Bert van Dam's Pneumatic Insect

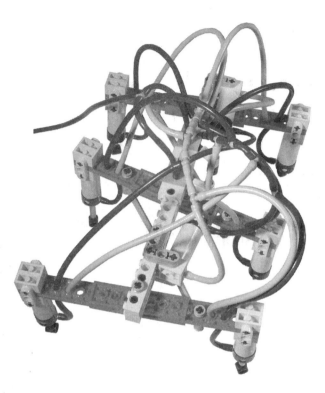

The complicated tubing hides the same basic circuit shown in Figure 10.16—
one of the control cylinders moves the three leg assemblies forward and backward, while the other moves the legs up and down. These are made of six
cylinders, split into two groups of three, controlled by the same valve. Each group
has a leg in a central position on one side, and one leg front and one leg rear on
the other side (see Figure 10.19).

Figure 10.19 Leg Connection Scheme for the Pneumatic Insect

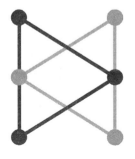

Though rather complicated to build, and more academic an example than practical, Van Dam's insect is quite amazing to see in action.

Using the same principle, it's possible to build a true pneumatic engine, where the push of the cylinders is converted into rotary motion exactly like in steam engines. Figure 10.20 shows our implementation of C.S.Soh's pneumatic engine. The key points are:

- Each cylinder has a dead point in its cycle, when it is either fully extended or retracted. In this position the cylinder is not able to perform any work, as its push/pull force cannot be converted into rotary motion. This happens because the two connection points of the cylinder (on the chassis and the wheel) and the fulcrum of the wheel align along the same line. For this reason, a pneumatic engine with a single cylinder would not work. The addition of a second cylinder solves the problem: You must mount it with a difference of 90° in its phase against the first one, so when one reaches a dead point, the other is at mid-stroke.

- The phasing of the valves is very important: You must take care to position them precisely, otherwise your engine won't work. Mount the wheels on the axles in such a way as to align one of their holes with the holes on the cams. Attach the liftarms to that hole with a gray pin. Connect the tubing exactly like that shown in Figure 10.20.

Figure 10.20 Soh's Pneumatic Engine

Pneumatic engines are capable of high torque, but due to their intrinsic friction are not suitable for high-speed applications. Most of the friction comes from the cylinders themselves, which, in order to be airtight, are a bit stiff to move.

Generally speaking, a vehicle moved by this engine, and supplied by an onboard compressor, is not very efficient. But it's indeed fun to see in action and might have its special uses, too (see Part III).

Summary

Beyond the fascinating sight of all those tubes, and the dramatic hissing of the air coming out of the valves, pneumatics have their practical strong points. In this chapter, you reviewed some basic concepts about the properties of gases, and learned how to exploit these when building your robots. Cylinders are definitely a better choice than electric motors for performing particular tasks, and, most significantly, have the capability to grab objects and create linear motion.

Electric compressors can provide a constant airflow to supply your cylinders, and can be used to control this flow from the RCX. Unfortunately, interfacing pneumatics to the RCX is not so simple, and requires a bulky assembly that includes an electric motor and some gearing. Perhaps in the future, the LEGO Company will produce a smart and compact interface able to control many valves from a single output port.

Pneumatics also offer the opportunity to implement simple automation based on cyclical operation, as we showed in the six-legged walker with its pneumatic engine.

Chapter 11

Finding and Grabbing Objects

Solutions in this chapter:

- Operating Hands and Grabbers
- Understanding Degrees of Freedom
- Finding Objects

Introduction

It's always great fun and very satisfying to see your robot pick things up from the ground, or take an object when you offer it. In this chapter, we'll illustrate some ways to build arms, hands, clamps, pliers, and other tools to grab and handle objects. One of the basic measurements of movement we'll explore is the *degree of freedom* (DOF), or the number of directions in which an object (like a robotic arm) has a range of motion. In the last part of the chapter, we'll show you methods by which your robot can find the objects, the most challenging part of the job.

Operating Hands and Grabbers

In Chapter 10, we illustrated that pneumatic cylinders are generally the ideal choice to make grabbing devices, or *grippers*. Unfortunately, pneumatics is not always a possible option. You might not have LEGO pneumatic parts, or you don't have room on your robot to fit a pneumatic compressor plus a pressure switch and some motor-driven valve switches. We've seen that RCX-to-pneumatics interfaces are rather cumbersome. So you must fall back on good old electric motors to drive your gripper.

The problem with motors is not opening or closing the hand, it's in getting the hand to apply a continuous pressure on the object to prevent it from falling. This means you cannot only position the fingers around it, you must also exert a force that tightens around the object even though you are not moving the fingers anymore. We have explained in Chapters 2 and 3 that if there's one thing that damages electric motors it is having them stalled, or rather having them powered but their movements blocked. For this reason, you cannot simply keep a motor turned on as the hand holds the object, you must employ a trick to prevent the motor from being permanently damaged. When you know you're going to handle a soft object that has some intrinsic elasticity, you can sometimes simply stop the motor and let the friction among gears keep the fingers against it. You can see a simple example of this in Figure 11.1, with an asymmetrical hand designed to grab sponge balls. The worm gear that drives the fingers prevents them from releasing the ball when the motor is not powered. Recall, from Chapter 2, that the worm gear is a one-way gear: It can turn a meshing gear but cannot be turned by it.

Figure 11.1 A Simple Hand Operated with a Worm Gear

Figure 11.2 shows a different design, where the rotary motion from the motor gets converted into linear motion through a worm gear and two translating axles. It's this motion that operates the movable fingers of a small hand. This mechanism is based on Leo Dorst's Electric Piston: Two half-bushings mesh the teeth of a worm gear; when the worm gear rotates, the bushings get pushed or pulled, and the axles where they are mounted move accordingly. Dorst's solution solves the problem of converting rotary to linear motion using a very compact scheme.

Figure 11.2 This Small Hand Uses Linear Motion

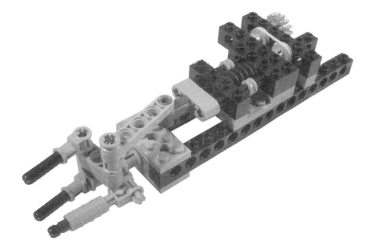

This approach doesn't work when your robotic hand is expected to handle rigid objects or ones of unknown shape and consistency. In these cases, you must introduce some elasticity into your system. Recall from Chapter 2 that you can use a pulley-belt setup to keep the motor running with no harm done even if the system gets blocked. Figure 11.3 shows a simple hand based on this principle: When you turn the motor on, the hand moves until it encounters enough resistance that the belt slips. While you keep the motor on, the belt transmits some force to the fingers and they hold the object. As soon as you stop the motor, however, the pressure of the finger drops and the object is released.

Figure 11.3 Running the Motor to Hold the Object

Even though it works, this solution is not very elegant because you're compelled to keep the motor running the entire time you want to hold the object. We suggest you use this system only if the robot must hold the object a very short time. In all other cases, you need something more reliable.

We've repeatedly said that pneumatic cylinders are your best choice in this field, but let's analyze what makes them so good to see if we can learn something and replicate the same behavior. A pneumatic cylinder can be considered a two-state system: The cylinder is either extended or retracted. (We are deliberately ignoring that you can somehow manually stop the cylinder in an intermediate position, centering the switch, and assuming that the switch is either in one of its extreme positions.) If something prevents the cylinder from actually reaching one of these states, it can, however, continue to push in that direction. Its natural behavior is to move until it finds resistance that balances its inner pressure. This pressure is what keeps the fingers applying a force to the object, thus making your robotic hand hold it firmly.

The point now is to replicate this behavior in a nonpneumatic device. Is it possible? Yes. Figure 11.4 contains an example of a simple *bi-stable* system, called bi-stable because it has two default states, two possible rest positions which it tends to go to. A rubber band forces the liftarm to stay against one of the two black pegs, either in A or in B. If you move the liftarm slightly from the peg and then release it, it goes back against the peg. If you move it a bit more and pass the midpoint between A and B, it goes to the other peg. You need to provide only enough force to make the system switch from one to the other; the rubber band will do the rest.

Figure 11.4 A Simple Bi-Stable Mechanism

Applying this principle, we designed the pliers shown in Figure 11.5, which are suitable for grabbing very small objects like a 1 x 2 brick (seen at the bottom of the Figure between the two plates). To actually use them in a robot you must add a motor that, through brief impulses, pushes the pliers into their open or closed states. As usual, you would probably involve a belt or a clutch gear to make the timing of the motor not critical.

The same approach can be used for larger and more complex hands, like the one shown in Figure 11.6, where the bi-stable mechanism has been placed on intermediate gearing.

Transferring Motion Using Tubing

In discussing the advantages of pneumatics when grabbing objects, we must also mention that tubing provides a simple way to keep bulky things far from the movable parts. Compare the simplicity of the pliers in Figure 11.7 with the complex gearings of the previous examples. The difference is dramatic.

Figure 11.5 Bi-Stable Pliers

Figure 11.6 A Bi-Stable Large Hand

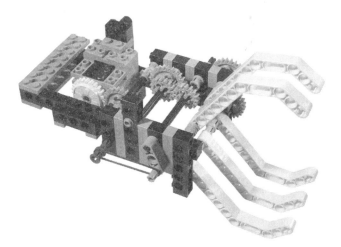

Figure 11.7 Pneumatics Helps in Making Essential and Clean Assemblies

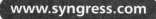

The flex system we briefly described in Chapter 9 has similar properties, allowing you to transfer motion to distant parts. Our robot, Cinque, features a small operating hand based on this technique (see Figure 11.8). A pair of opposing rubber bands introduce a degree of elasticity into the system, and help the fingers return to their default setting once the hand comes to rest in its open position.

Figure 11.8 The Flex System Helps in Making Lightweight Hands

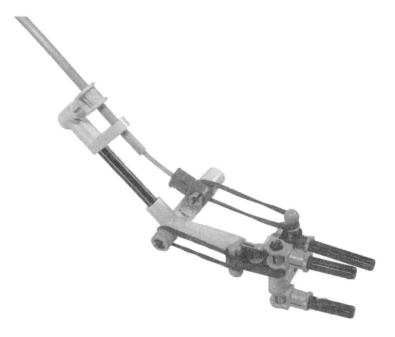

Understanding Degrees of Freedom

If you look carefully at your hands, you'll discover they are an incredible piece of machinery, capable of handling a wide array of objects of every size and shape. Just think about the long list of verbs describing all the things hands can do: grab, handle, hold, take, squeeze, grip, point, pinch, shake, roll, press, grasp, push, pull and those are only a few of the terms. Where does all this versatility come from?

Observe a finger while you move it, you notice four independent movements: three for the joints—from the finger tip to the hand—that let you bend the finger, and a fourth that allows for slight left-right motion where the finger joins the hand. Multiply this by five (for a hand's five fingers) and add the mobility given by the wrist, and 25 movements or so come to mind, which, in turn, lead to a huge number of combinations and configurations. This is what makes your hand able to conform to the shape of the object you want to handle. To complete the picture, consider that you can control the strength of each muscle so finely that you can pick up a delicate wine glass without damaging it, yet so firmly grip a baseball bat that you can send a ball over the right field wall.

Every independent movement represents a degree of freedom (DOF), something that can happen without affecting and being affected by other movements in the same device. Our previous examples were very simple mechanisms with just one degree of freedom, being that all the possible positions of the "fingers" were determined by a single motor or pneumatic cylinder. The DOF concept helps you understand in terms of numbers why those simple hands diverged so widely from the flexibility a human hand has.

Obviously, you cannot aim at making a robotic hand with 25 degrees of freedom using your MINDSTORMS kit. Each degree of freedom will typically require a dedicated motor or pneumatic cylinder, and this puts the task out of reach. You should stay with something much simpler and consequently reduce the range of objects your mechanical hand will be able to grab. This is sometimes limiting, but in many situations you will know in advance the type and shape of objects your robot will be expected to handle, making your task less demanding. In contests that involve collecting things, for example, your robot usually will deal with very specific objects like soda cans, small LEGO cubes or marbles, and because of this you can design it to target those types of objects.

It is possible, however, to build more versatile hands with more degrees of freedom. Figure 11.9 shows a 3 degrees of freedom pneumatic finger. This is a nice design, but it's a pity it requires all three ports of your RCX to be fully controlled. How could you control more than a finger if you are already out of ports? To make the system simpler, though still useful, you can connect all the cylinders together. (You won't be able to move a single segment of the finger by itself, but the finger can still adapt well to the shape of many different objects.) This is the technique we used in the three-finger pneumatic hand shown in Figures 11.10 and 11.11, which is controlled by a single valve switch.

Figure 11.9 A Three Degrees of Freedom Pneumatic Finger

Figure 11.10 A Three-Finger Pneumatic Hand

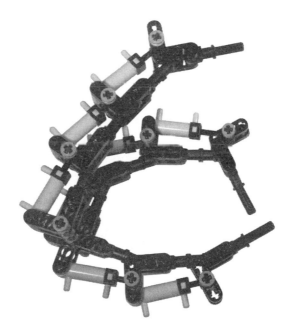

Figure 11.11 The Three-Finger Pneumatic Hand with Complete Tubing

The degrees of freedom concept applies not only to hands but to any mechanical device. The arm of Figure 11.12, taken from our R2-D2 styled robot "Otto," has two degrees of freedom: A large cylinder extends the arm from the body of the robot, and a small one operates the hand.

Figure 11.12 The Robotic Arm from Our "Otto" Robot

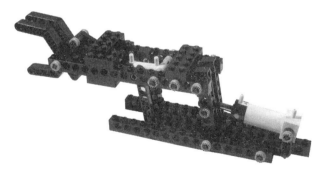

Generally speaking, locating a point in a plane requires two DOFs, while locating a point in space requires three. There are many examples of 2 DOF- and 3 DOF-mechanisms in everyday objects: An ink-jet printer has two DOFs, one corresponding to the head movement and the other to the paper feeding. A construction crane is an example of a machine with three DOFs: The hook can go up and down, it's attached to a carriage that moves back and forth along the boom, and, finally, the boom can rotate. With the three output ports of your RCX, you can drive a robotic arm that addresses any point inside a delimited space, called the *operating envelope*, exactly like the crane of the previous example. If you also want to pick up and drop objects, you would need another port, or use some of the tricks from Chapter 9 to control more than a DOF with a single motor.

Finding Objects

Building robotic arms and hands is the easy part of the job, the hard part is finding the objects to grab. We will skip the case where your robot *knows* the position of one of the objects, because this brings into play a general navigation problem we'll discuss in Chapter 13. So, for the time being, we'll stick with the fact that the robot knows nothing about the location of the object.

As we explained when talking about bumpers in Chapter 4, navigation in real environments is quite a tough task, and distinguishing a specific object from others makes things much harder. So the second assumption we make here is that you know what kind of object you're expected to handle, as well as all the details of the environment where your robot moves (typically an artificial one prepared for the task). You might think that we are introducing too many simplifications here, but even in these conditions, the task remains quite hard. It's very important that you progress in short steps. The most common mistake of beginning builders is to start out with goals too difficult for their robots, where mechanical and programming difficulties add to navigation problems. As a general approach, we suggest you apply the "divide and conquer" strategy and solve the problems one by one.

Let's make an example: A simple variation on line following that might involve removing objects placed along the path. A very simple bumper is probably enough to detect objects. The arm will be more or less sophisticated depending on whether you have to collect them or just move them out of the way.

In wider environments, things become trickier. Imagine you have to find things in a delimited space with no walls. (How could a space be delimited without having walls? By using different colors on the floor and reading them with a light sensor facing down!) You can still use a bumper, and make your robot move around at random or follow some kind of scheme. Depending on whether you are participating a contest with specific rules, you could make this approach more efficient using a sort of funnel to convey the objects against the bumper, or some long antennas to help you detect contacts in a wider area. The robot of

Figure 11.13 was designed to find small LEGO cubes during a contest, and takes advantage of the fact that the height of the object is known precisely enabling us to detect the cubes with a top bumper.

Figure 11.13 StudWhite, a LEGO Cubes Finder

In other situations, you can apply the proximity detection technique, either with standard LEGO components as described in Chapter 4, or with custom IRPD sensors like the one shown in Chapter 9. Let's go back to the example where there are no walls. You can use proximity to "see" the objects, maybe improving final detection with a bumper as in previous scenarios. And if there *are* walls? Well, you'll need a way to distinguish the objects from the walls.

The easiest approach is to rely again on the shape of the object. Usually the walls are taller then the soda cans or marbles you have to find, so you can prepare two bumpers at different heights and see which one closes to decide what your robot ran into. The same works with proximity detection: Placing two sensors at different heights will tell you whether you've found a soda can or the wall (Figure 11.14). Be careful though… Two or more active custom proximity sensors, the kind that emit their own IR beam, can interfere with each other, resulting in unreliable readings. Instead of receiving back just the IR light that they emit, each one will also receive the IR light emitted by their brother. To avoid this problem, you have to write your software to make them active one at a time. This can be achieved configuring them as passive sensors (for example, as touch sensors), so they don't receive any power, and consequently don't emit any IR beam. Your program will configure them as active sensors just before performing the reading, and will change them to passive sensors again afterward.

Figure 11.14 Using Two IRPDs to Distinguish Objects from Walls

A different case is when you want to manually trigger your robot to grab or release objects. This is very easy to implement with a touch sensor, a push button that you press when you want your robot to open or close its hand. Proximity detection makes your robot even more impressive to see in action. You can, for instance, build a robot that navigates the room, and that, when you offer it an object, stops to grab it. This technique is a bit tricky to use if your robot is expected to navigate a room with walls and other obstacles, because it won't be able to tell what triggered its proximity detection, unless you have a custom sensor that returns some relative or absolute measurement of the distance. In this case, you can continuously monitor the distance and interpret a sudden radical change in its movement as a request to grab or release objects.

Summary

Designing a good robotic hand or arm is more of an art than a technique. There are indeed technical issues when it comes to gearing and pneumatics that you must know and consider to successfully position the grippers or hands, apply the right amount of pressure, troubleshoot the elasticity of the object to be grabbed, and not allow your robot to drop the ball (or object rather). Even then, there's still a lot of space for good intuitions and heavy prototyping. You can choose pneumatic or nonpneumatic approaches, design for different degrees of freedom in your gripping arm, use a flex system with tubing for lightweight designs, and create solutions that reserve ports for additional functions.

To make an easy start, target your first projects around a specific type of object, then progress to more versatile grabbers only when you feel experienced and confident enough to meet the challenge.

We also explained that finding the object is the hardest part of the job, but there are cases where you can use a random search pattern, or where the object sits on the robot's path, as in the line following example.

Chapter 12

Doing the Math

Solutions in this chapter:

- **Multiplying and Dividing**
- **Averaging Data**
- **Using Interpolation**
- **Understanding Hysteresis**

Introduction

You may be surprised to find a chapter about mathematics in a book aimed at explaining building techniques. However, just as one can't put programming aside totally, so too we cannot neglect an introduction to some basic mathematical techniques. As we've explained, robotics involves many different disciplines, and it's almost impossible to design a robot without considering its programming issues together with the mechanical aspects. For this reason, some of the projects we are going to describe in Part II of the book include sample code, and we want to provide here the basic foundations for the math you will find in that code. Don't worry, the math we'll discuss in this chapter doesn't require anything more sophisticated than the four basic operations of adding, subtracting, multiplying, and dividing. The first section, about multiplying and dividing, explains in brief how computers deal with integer numbers, focusing on the RCX in particular. This topic is very important, because if you are not familiar with the logic behind computer math you are bound to run into some unwanted results, which will make your robot behave in unexpected ways.

The three subsequent sections deal with *averages*, *interpolation*, and *hysteresis*. Though they are not essential, you should consider learning these basic mathematical techniques, because they can make your robot more effective while at the same time keep its programming code simpler. Averages cover those cases where you want a single number to represent a sequence of values. School grades are a good example of this: They are often averaged to express the results of students with a single value (as in a grade point average). Robotics can benefit from averages on many occasions, especially those situations where you don't want to put too much importance on a single reading from a sensor, but rather observe the tendency shown by a group of spaced readings.

Interpolation deals with the estimating, in numerical terms, of the value of an unknown quantity that lies between two known values. Everyday life is full of practical examples—when the minute hand on your watch is between the Three and Four marks, you interpolate that data and deduce that it means, let's say, eighteen minutes. When a car's gas gauge reads half a tank, and you know that with the full tank the car can cover about four hundred miles, you make the assessment that the car can currently travel approximately two hundred miles before needing refueling. Similarly in robotics, you will benefit from interpolation when you want to estimate the time you have to operate a motor in order to set a mechanism in a specific position, or when you want to interpret readings from a sensor that fall between values corresponding to known situations.

The last tool we are going to explore is *hysteresis*. Hysteresis defines the property of a variable for which its transition from state A to state B follows different rules than its transition from state B to state A. Hysteresis is also a programmed behavior in many automatic control devices, because it can improve the efficiency of the system, and it's this facet that interests us. Think of hysteresis as being similar to the word "tolerance," describing, in other words, the amount of fluctuation you allow your system before undertaking a corrective action. The hysteresis section of the chapter will explain how and why you might add hysteresis to the behavior of your robots.

Multiplying and Dividing

If you are not an experienced programmer, first of all we want to warn you that in the world of computers, mathematics may be a bit different from what you've been taught in school, and some expressions may not result in what you expect. The math you need to know to program the small RCX is no exception.

Computers are generally very good at dealing with *integer* numbers, that is, whole numbers (1, 2, 3...) with the addition of zero and negative whole numbers. In Chapter 6, we introduced *variables*, and explained that variables are like boxes aimed at containing numbers. An *integer variable* is a variable that can contain an integer number. What we didn't say in Chapter 6 is that variables put limits on the size of the numbers you can store in them, the same way that real boxes can contain only objects that fit inside. You must know and respect these limits, otherwise your calculations will lead to unexpected results. If you try to pour more water in a glass than what it can contain, the exceeding water will overflow. The same happens to variables if you try to assign them a number that is greater than their capacity—the variable will only retain a part of it.

The firmware of the RCX has been designed to manipulate integer numbers in the range −32768 through 32767. This means that when using either RCX Code, NQC, or any other language based upon the LEGO firmware, you must keep the results of your calculations inside this range. This rule applies also to any intermediate result, and entails that you learn to be in control of your mathematics. If your numbers are outside this range, your calculations will return incorrect results and your robot will not perform as expected; in technical terms, this means you must know the *domain* of the numbers you are going to use. Multiplication and division, for different reasons, are the most likely to give you trouble.

Let's explain this statement with an example. You build a robot that mounts wheels with a circumference of 231mm. Attached to one wheel is a sensor geared to count 105 ticks per each turn of the wheel. Knowing that the sensor reads a count of 385, you want to compute the covered distance. Recall from Chapter 4 that the distance results from the circumference of the wheel multiplied by the number of counts and divided by the counts per turn:

231 x 385 / 105 = 847

This simple expression has obviously only one proper result: 847. But if you try to compute it on your RCX, you will find you can *not* get that result. If you perform the multiplication first, that is, if the expression were written as follows:

(231 x 385) / 105

you get 222! If you try and change the order of the operations this way:

231 x (385 / 105)

you get 693, which is closer but still wrong! What happened? In the first case, the result of performing the multiplication first (88,935) was outside the upper limit of the allowed range, which is only 32,767. The RCX couldn't handle it properly and this led to an unexpected result. In the second case, in performing the division operation first, you faced a different problem: The RCX handles only integers, which cannot represent fractions or decimal numbers; the result from 385 / 105 should have been 3 2/3, or 3.66666..., but the processor truncated it to 3 and this explains the result you got.

Unfortunately, there is no general solution to this problem. A dedicated branch of mathematics, called *numerical analysis*, studies how to limit the side effects of mathematical operations on computers and quantify the expected errors and their propagation along calculations. Numerical analysis teaches that the same error can be expressed in two ways: *absolute errors* and *relative errors*. An absolute error is simply the difference between the result you get and the true value. For example, 4355 / 4 should result in 1088.75; the RCX truncates it to 1088, and the absolute error is 1088.75 − 1088 = 0.75. The division of 7 by 4 leads to the same absolute error: The right result is 1.75, it gets truncated to 1, and the absolute error is again 0.75. To express an error in a relative way, you divide the absolute error by the number to which it refers. Usually, relative errors gets converted into *percentage errors* by multiplying them by one hundred. The percentage errors of our previous examples are quite different one from the other: 0.07 percent for the first one (0.75 / 1088.75 x 100) and an impressive 42.85 percent error for the latter (0.75 / 1.75 x 100)! Here are some useful tips to remember from this complex study:

- You have seen that integer division will result in a certain loss of precision when decimals get truncated. Generally speaking you should perform divisions as the *last step* of an expression. Thus the form (A x B) / C is better than A x (B / C), and (A + B) / C is better than its equivalent A / C + B / C.

- While integer divisions lead to small but predictable errors, operations that go off-range (called *overflows* and *underflows*) result in gross mistakes (as you discovered in the example where we multiplied 231 by 385). You must avoid them at all costs. We said that the form (A x B) / C is better than A x (B / C), but *only* if you're sure A x B doesn't overflow the established range!

- When dividing, the larger the dividend over the divisor, the smaller the relative error. This is another reason (A x B) / C is better than A x (B / C): The first multiplication makes the dividend bigger.

- Prescaling values to relocate them in a different range is sometimes a good option, especially if you can do so without a loss in accuracy. For example, if you are dealing with raw values coming from a light sensor, they will likely be in the range of 550 to 850. In the event you had to multiply them with other numbers, you could subtract 500 from all your readings to move them down into the range of 50 to 350.

Designing & Planning...

Floating-Point Numbers

If you are familiar with computer programming, you probably know that many languages support another common numeric format: *floating-point*. The internal representation of a floating-point number is made up of two values, a *mantissa* and an *exponent*, and corresponds to the number that results multiplying the mantissa by a conventional *base* raised to the exponent. This technique allows floating-point variables to handle numbers in a very wide domain.

Continued

Up to this point, we deliberately omitted talking about them for two reasons. First, the RCX firmware doesn't support floating-points (currently only legOS and leJOS can handle them), and second, they don't result, by themselves, in a greater precision. As for integers, their precision is limited to the number of bits used to map them in memory.

We admit that they do provide a convenient way to represent values from a wider range then integers, with fewer concerns about overflows and truncations, but in robotics it's really possible to face most situations with only integer math.

Averaging Data

There are situations when you may prefer that your robot base its decisions not on a single sensor reading but on a *group* of them, to achieve more stable behavior. For example, if your robot has to navigate a pad made up of colored areas rather than just black and white, you would want it to change its route only when it finds a different color, ignoring transition areas between two adjacent colors (or even dirt particles that could be "read" by accident).

Another case is when you want to measure ambient light, ignoring strong lights and shadows. *Averaging* provides a simple way to solve these problems.

Simple Averages

You're probably already familiar with the simple average, the result of adding two or more values and dividing the sum by the number of addends. Let's say you read three consecutive light values of 65, 68, and 59, their simple average would be:

(65 + 68 + 59) / 3 = 64

which is expressed in the following formula:

$A = (V_1 + V_2 + \ldots + V_n) / n$

The main property of the average, what actually makes it useful to our purpose, is that it smoothes single peak values. The larger the amount of data averaged, the smaller the effect of a single value on its result. Look at the following sequence:

60, 61, 59, 58, 60, 62, 61, 75

The first seven values fall in the range of 58 to 62, while the eighth one stands out with a 75. The simple average of this series is 62, thus you see that this last reading doesn't have a strong influence (Figure 12.1).

Figure 12.1 How Averaging Smoothes Peaks and Valleys in the Data

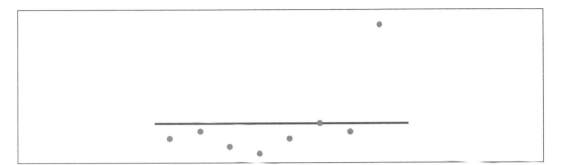

In your practical applications, you won't average all the readings of a sensor, usually just the last n ones. It is like saying you want to benefit from the smoothing property of an average, but only want to consider more recent data because older readings are no longer relevant.

Every time you read a new value, you discard the oldest one with a technique called the *moving average*. It's also known as the boxcar average. Computing a moving average in a program requires you to keep all the last n values stored in variables, then properly initialize them before the actual executions begins. Think of a sequence of sensor values in a long line. Your "boxcar" is another piece of paper with a rectangular cutout in it, and you can see exactly n consecutive values at any one time. As you move the boxcar along the line of sensor values, you average the readings you see in the cutout. It is clear that as you move the boxcar by one value from left to right along the line, the leftmost value drops off and the rightmost value can be added to the total for the average.

Going back to the series from our previous example, let's build a moving average for three values. You need the first three numbers to start: 60, 61, and 59. Their average is (60 + 61 + 59) / 3 = 60. When you receive a new value from your sensor, you discard the oldest one (60) and add the newest (58). The average now becomes (61 + 59 + 58) / 3 = 59.333... Figure 12.2 shows what happens to the moving average for three values applied to all the values of the example.

When raw data shows a trend, moving averages acknowledge this trend with a "lag." If the data increases, the average will increase as well, but at a slower pace. The higher the number of values used to compile the average, the longer the lag.

Suppose you want to use a moving average for three values in a program. Your NQC code could be as follows:

```
int ave, v1, v2, v3;


v2 = SENSOR_1;
v3 = SENSOR_1;


while (true)
{
  v1 = v2;
  v2 = v3;
  v3 = SENSOR_1;
  ave = (v1+v2+v3) / 3;
  // other instructions...
}
```

Figure 12.2 A Moving Average for Three Values

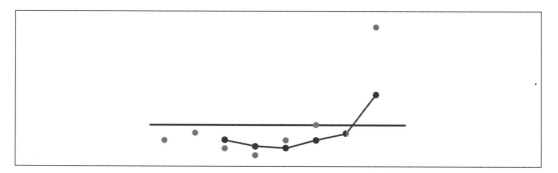

Note the mechanism in this code that drops the oldest value (v1), replacing it with the subsequent one (v2), and that shifts all the values until the last one is replaced with a fresh reading from the sensor (in v3). The average can be computed through a series of additions and a division.

When the number of reading being averaged is large, you can make your code more efficient using *arrays*, adopting a trick to improve the computation time and keep the number of operations to a minimum. If you followed the description of the boxcar cutout as it moved along the line, you would realize that the total of the values being averaged did not have to be calculated every time. We just need to subtract the leftmost value, and add the rightmost value to get the new total!

A circular pointer, for example, can be used to address a single element of the array to substitute, without shifting all the others down. The number of additions, meanwhile, can be drastically decreased keeping the total in a variable, subtracting the value that exits, and adding the entering one. The following NQC code provides an example of how you can implement this technique:

```
#define SIZE 3
int v[SIZE],i,sum,ave;

// initialize the array
sum = 0;
for (i=0;i<SIZE-1;i++)
{
  v[i] = SENSOR_1;
  sum += v[i];
}

// first element to assign is the last of the array
i=SIZE-1;
v[i]=0;

// compute moving average
while (true)
{
  sum -= v[i]; //
  v[i] = SENSOR_1;
  sum += v[i];
  ave = sum / SIZE;
  i = (i+1) % SIZE;
  // other instructions...
}
```

The constant SIZE defines the number of values you want to use in your moving average. In this example, it is set to 3, but you can change it to a different number. The statement **int** declares the variables; **v[SIZE]** means that the variable **v** is an *array*, a container with multiple "boxes" rather than a single "box." Each *element* of the array works exactly like a simple variable, and can be addressed specifying its position in the array. Array elements are numbered starting from 0, thus in an array with 3 elements they are numbered 0, 1, and 2. For example, the second element of the array **v** is **v[1]**.

This program starts initializing the array with readings from the sensor. It uses the **for** control statement to loop SIZE-1 times, at the same time incrementing the **i** variable from 0 to SIZE-1. Inside the loop, you assign readings from the sensor to the first SIZE-1 elements of the array. At the same time, you add those values to the **sum** variable. Supposing that the first readings are 72 and 74, after initialization **v[0]** contains 72, **v[1]** contains 74, and **sum** contains 146. The initialization process ends assigning to the variable **i** the number of the first array element to replace, which corresponds to SIZE-1, which is 2 in our example.

Let's see what happens inside the loop that computes the moving average. Before reading a new value from the sensor, we remove the oldest value from **sum**. The first time **i** is 2 and **v[i]**, that is **v[2]**, is 0, thus **sum** remains unchanged. v[i] receives a new reading from the sensor and this is added to **sum**, too. Supposing it is 75, **sum** now contains 146 + 75 = 221. Now you can compute the average **ave**, which results in 221 / 3 = 73.666…, and which is truncated to 73. The following instruction prepares the pointer **i** to the address of the next element that will be replaced. The symbol **%** in NQC corresponds to the *modulo* operator, which returns the remainder of the division. This is what we call a *circular pointer*, because the expression keeps the value of **i** in the range from 0 to SIZE-1. It is equivalent to the code:

```
i = i+1;
if (i==SIZE) i=0;
```

which resets **i** to 0 when it reaches the upper bound. The resulting effect is that **i** cycles among the values 0, 1, and 2.

During the second loop **i** is 0, so **sum** gets decreased to **v[0]**, that is 72, and counts 221 − 72 = 149. **v[0]** is now assigned a new reading, for example 73, and **sum** becomes equal to 149 + 73 = 222. The average results 222 / 3 = 74, and **i** is incremented to 1. Then the cycle starts again, and it's time for **v[1]** to be replaced with a new value.

This program is definitely much more complicated than the previous one, but has the advantage of being more general: It can compute moving averages for any number of values by just changing the SIZE constant. The only limit to the maximum value of SIZE is the total number of variables allowed by the LEGO firmware, which is 32. Each array element counts as a variable.

Weighted Averages

We explained that simple averages have the property of smoothing peaks and valleys in your data. Sometimes, though, you want to smooth data to reduce the effect of single readings, yet at the same time put more importance on recent values than older ones. In fact, the more recent the data, the more representative the possible trend in the readings.

Let's suppose your robot is navigating a black and white pad, and that it's crossing the border between the two areas. The last three readings of its light sensor are 60, 62, and 67, which result in a simple average of 63. Can you tell the difference between that situation and one where the readings are 66, 64, and 59 using just the simple average? You can't, because both series have the same average. However, there's an evident diversity between the two cases—in the first, the readings are increasing, in the latter they are decreasing but the simple average cannot separate them. In this case, you need a *weighted* average, that is, an average where the single values get multiplied by a factor that represents their importance.

The general formula is:

$$A = (V_1 \times W_1 + V_2 \times W_2 + ... + V_n \times W_n) / (W_1 + W_2 + ... + W_n)$$

Suppose you want to give a weight of 1 to the oldest of three readings, 2 to the middle, and 4 to the latest one. Let's apply the formula to the series of our example:

$$(60 \times 1 + 62 \times 2 + 67 \times 4) / (1 + 2 + 4) = 64.57$$
$$(66 \times 1 + 64 \times 2 + 59 \times 4) / (1 + 2 + 4) = 61.43$$

You notice that the results are very different in the two cases: The weighted average reflects the trend of the data. For this reason, weighted averages seem ideal in many cases. They allow you to balance multiple readings, at the same time taking more recent ones into greater consideration. Unfortunately, they are memory- and time-consuming when computed by a program, especially when you want to use a large number of values.

Now, there is a particular class of weighted averages that can be of help, providing a simple and efficient way of storing historical readings and calculating new values. They rely on a method called *exponential smoothing*. (Don't let the name frighten you!)

The trick is simple: You take the new reading and the previous average value, and combine these into a new average value using two weights that together represent 100 percent. For example, you can take 40 percent of the new reading and 60 percent of the previous average, or instead take only 10 percent of the new reading and 90 percent of the previous average. The less weight you put on the new value, the more stable and slow to change the average will be.

The general equation for exponential smoothing is expressed as follows:

$$A_n = (V_n \times W_1 + A_{n-1} \times W_2) / (W_1 + W_2)$$

You can choose W_1 and W_2 to add up to 100, so that you can easily read them as a percentage. For example:

$$A_n = (V_n \times 20 + A_{n-1} \times 80) / 100$$

Let's apply this formula to the series of the previous example. The first number in the first series was 60. There is no previous value for the average, so we simply take this number:

$$A_1 = 60$$

When the next reading (62) arrives, you compute a new value for the average using the whole formula:

$$A_2 = (62 \times 20 + 60 \times 80) / 100 = 60.4$$

Then you apply the rule again for the third value:

$$A_3 = (67 \times 20 + 60.4 \times 80) / 100 = 61.72$$

The result tells you that the average is *slowly* acknowledging the trend in the data. This happens because the last reading counts only for 20 percent, while 80 percent comes from the previous value. If you want to make your average more reactive to recent values, you must increase the weight of the last factor. Let's see what happens by changing the 20 percent to 60 percent:

$$A_1 = 60$$
$$A_2 = (62 \times 60 + 60 \times 40) / 100 = 61.2$$
$$A_3 = (67 \times 60 + 61.2 \times 40) / 100 = 64.68$$

You notice that the formula is still smoothing the values, but gives much more importance to recent values. One of the advantages of exponential smoothing is that it is very easy to implement. The following is an example of NQC code:

```
int ave;

// initialize the average
ave = SENSOR_1;

// compute average
while (true)
{
  ave = (SENSOR_1 * 20 + ave * 80) / 100;
  // other instructions...
}
```

Simple, isn't it? You could be tempted to *reduce* the mathematical expression, but be careful, remember what we said about multiplying and dividing integer numbers. These are okay:

```
ave = (SENSOR_1 * 2 + ave * 8) / 10;
ave = (SENSOR_1 + ave * 4) / 5;
```

But this, though mathematically equivalent, leads to a worse approximation:

```
ave = SENSOR_1 / 5 + ave * 4 / 5;
```

Designing & Planning...

Exponential Smoothing

Those of you with a gift for math might be interested in understanding where exponential smoothing got its name. Let's try to analyze the equation:

$$A_n = (V_n \times W_1 + A_{n-1} \times W_2) / (W_1 + W_2)$$

We can rewrite the weights W1 and W2 as fractions: $w_1 = W_1 / (W_1 + W_2)$ and $w_2 = W_2 / (W_1 + W_2)$, where w_1 and w_2 result in the range of 0 to 1. As $w_1 + w_2 = 1$, we can substitute w_2 with $(1 - w_1)$. Our equation then becomes:

$$A_n = V_n \times w_1 + A_{n-1} \times (1 - w_1)$$

Continued

Expanding the term A_{n-1} we get:

$$A_{n-1} = V_{n-1} \times w_1 + A_{n-2} \times (1 - w_1)$$

and substituting in the previous:

$$A_n = V_n \times w_1 + V_{n-1} \times w_1 \times (1 - w_1) + A_{n-2} \times (1 - w_1)^2$$

Continuing this expansion, we get the general form:

$$A_n = V_n \times w_1 + V_{n-1} \times w_1 \ (1 - w_1) + V_{n-2} \times w_1 \times (1 - w_1)^2$$
$$+ \ldots + V_{n-m} \times w_1 \times (1 - w_1)^m + \ldots + V_1 \times w_1 \times (1 - w_1)^n$$

This average is thus equivalent to an average of all the values, where the older they are the more they get smoothed by the exponential term $(1 - w_1)^m$. The term $(1 - w_1)$ being less than zero, means the higher the exponent, the smaller the result.

Using Interpolation

You've built a custom temperature sensor that returns a raw value of 200 at 0°C and 450 at 50°C. What temperature corresponds to a raw value of 315? Your robotic crane takes 10 seconds to lift a load of 100g, and 13 seconds for 200g. How long will it take to lift 180g? To answer these and other similar questions, you would turn to *interpolation*, a class of mathematical tools designed to estimate values from known data.

Interpolation has a simple geometric interpretation: If you plot your known data as points on a graph, you can draw a line or a curve that connects them. You then use the points on the line to guess the values that fall inside your data. There are many kinds of interpolation, that is, you can use many different equations corresponding to any possible mathematical curve to interpolate your data. The simplest and most commonly used one is *linear interpolation*, for which you connect two points with a straight line, and this is what we are going to explain here (Figure 12.3).

Please be aware that many physical systems don't follow a linear behavior, so linear interpolation will not be the best choice for all situations. However, linear interpolation is usually fine even for nonlinear systems, if you can break the ranges down into almost linear sections.

Figure 12.3 Linear Interpolation

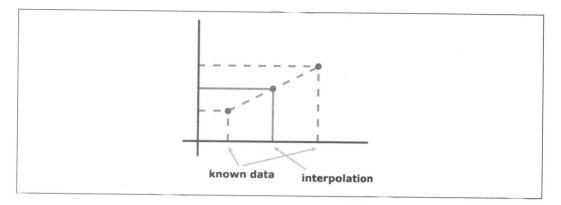

known data   interpolation

In following standard terminology, we will call the parameter we change the *independent variable*, and the one that results from the value of the first, the *dependent variable*. With a very traditional choice, we will use the letter X for the first and Y for the second. The general equation for linear interpolation is:

$$(Y - Y_a) / (Y_b - Y_a) = (X - X_a) / (X_b - X_a)$$

Where Y_a is the value of Y we measured for $X = X_a$ and Y_b the one for X_b. With some simple work we can isolate the Y and transform the previous equation into:

$$Y = (X - X_a) \times (Y_b - Y_a) / (X_b - X_a) + Y_a$$

This is very simple to use, and allows you to answer your question about the custom temperature sensor. The raw value is your independent variable X, the one you know. The terms of the problem are:

$X_a = 200 \; Y_a = 0$
$X_b = 450 \; Y_b = 50$
$X = 315 \; Y = ?$

We apply the formula and get:

$$Y = (315 - 200) \times (50 - 0) / (450 - 200) + 0 = 23$$

To make our formula a bit more practical to use, we can transform it again. We define:

$$m = (Y_b - Y_a) / (X_b - X_a)$$

If you are familiar with college math, you will recognize in **m** the slope of the straight line that connects two points. Now our equation becomes:

$$Y = m \times X - m \times X_a + Y_a$$

As s, X_a and Y_a are all known constants, we compute a new term b as:

$$b = Y_a - m \times X_a$$

so our final equation becomes:

$$Y = m \times X + b$$

This is the standard equation of a straight line in the Cartesian plane. Looking back to our previous example, you can now compute **m** and **b** for your temperature sensor:

$$m = (50 - 0) / (450 - 200) = 0.2$$
$$b = 0 - 0.2 \times 200 = -40$$
$$Y = 0.2 \times X - 40$$

You can confirm your previous result:

$$Y = 0.2 \times 315 - 40 = 23$$

Implementing this equation inside a program for the RCX will require that you convert the decimal value 0.2 into a multiplication and a division, this way:

```
temp = (raw * 2) / 10 - 40;
```

Interpolation is also a good tool when you want to relocate the output from a system in a different range of values. This is what the RCX firmware does when converting raw values from the light sensor into a percentage (see Chapter 4). You can do the same in your application. Suppose you want to change the way raw values from the light sensor get converted into a percentage. The LEGO firmware defines that 1022 converts to 0 percent and 322 to 100 percent, but this range is quite wide with regard to the readings you actually experience with the light sensor. Let's say you want to fix an arbitrary range of 900 converting to 0 percent and 500 converting to 100 percent, and this is what you get from the interpolation formula:

$$m = (0 - 100) / (900 - 500) = -0.25$$
$$b = 100 + 0.25 \times 500 = 225$$
$$Y = -0.25 \times X + 225$$

Multiplying by 0.25 is like dividing by 4, so we can write this expression in code as:

```
perc = - raw / 4 + 225;
```

Understanding Hysteresis

Hysteresis is actually more a physical than a mathematical concept. We say that a system has some hysteresis when it follows a path to go from state A to state B, and a different path when going *back* from state B to state A. Graphing the state of the system on a chart shows two different curves that connect the points A and B, one for going out and one for coming back (Figure 12.4).

Figure 12.4 Hysteresis in Physical Systems

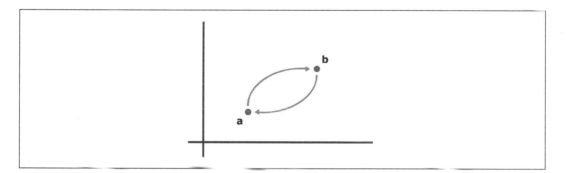

Hysteresis is a common property of many natural phenomena, magnetism above all, but our interest here is in introducing some hysteresis in our robotic programs. Why should you do it? First of all, let us say this is quite a common practice. In fact, many automation devices based on some kind of feedback have been equipped with artificial hysteresis.

A very handy example comes from the thermostat that controls the heating in your house. Imagine your heating system relies on a thermostat designed to maintain an exact temperature, and that during a cold winter you program your desired home temperature to 21°C (70°F). As soon as the ambient temperature goes below 21°C, the heating starts. In a few minutes the temperature reaches 21°C and heating stops, then a few minutes later starts again and so on all day long. The heater would turn off and on constantly as the temperature varies around that exact point. This approach is not the best one, because every start phase requires some time to bring the system to its maximum efficiency, just about when it gets stopped again. In introducing some hysteresis, the system can run smoother: We can let the temperature go down to 20.5°C, then heat up the house until it reaches 21.5°C. When the temperature in the house is 21°C, the heating can be either on or off, depending whether it's going from on to off or vice versa.

Hysteresis can reduce the number of corrective actions a system has to take, thus improving stability and efficiency at the price of some tolerance. Auto-pilots for boats and planes are another good example. Could you think of a task for your robots that could benefit from hysteresis? Line following is a good example.

In Chapter 4, in talking about light sensors, we explained that the best way to follow a strip on the floor is to actually follow one of its edges, the borderline between white and black. In that area, your sensor will read a **gray** value, some intermediate number between the white and black readings. Having chosen that value for **gray**, a robot with no hysteresis may correct left when the reading is greater than **gray** and right when reading is less than **gray**. To introduce some hysteresis, you can tell your robot to turn left when reading **gray+h** and right when reading **gray-h**, where h is a constant properly chosen to optimize the performance of your robot. There isn't a general rule valid for any system, you must find the optimal value for h by experimenting. Start with a value of about 1/6 or 1/8 of the total white-black difference; this way the interval **gray-h** to **gray+h** will cover 1/3 or 1/4 of the total range. Then start increasing or decreasing its value observing what happens to your robot, until you are happy with its behavior. You will discover that by reducing **h** your robot will narrow the range of its oscillations, but will perform more frequent corrections. Increasing **h**, on the other hand, will make your robot perform less corrections but with oscillations of larger amplitude (Figure 12.5).

Figure 12.5 How Hysteresis Affects Line Following

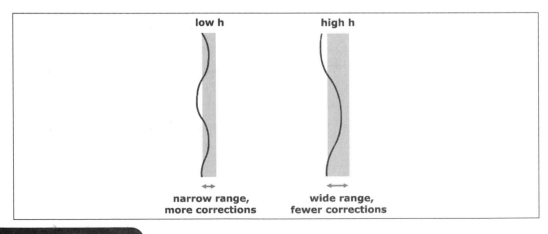

We suggest a simple experiment that will help you put these concepts into practice by building a real sensor setup that you can manipulate by hand to get a feeling of how the robot would behave. Write a simple program that plays tones to ask you to turn left or right. For example, it can beep high when you have to turn left and low to turn right. The NQC code that follows shows a possible implementation:

```
#define GRAY 50
#define H 3


task main()
{

  SetSensor(SENSOR_1,SENSOR_LIGHT);

  while (true)
  {
    if (SENSOR_1>GRAY+H)
      PlayTone(440,20);
    else if (SENSOR_1<GRAY-H)
      PlayTone(1760,20);
    Wait(20);
  }
}
```

Equip your RCX with a light sensor attached to input port 1 and you are ready to go. You should tune the value of the *GRAY* constant to what your sensor actually reads when placed exactly over the borderline between the white and the black. When the program runs, you can move the sensor toward or away from the line until you hear the beep that asks you to correct the direction. (Keep the sensor always at the same distance from the pad.) Experiment with different values for H to see how the accepted range of readings gets wider or narrower.

If you keep a pencil in your hand together with the light sensor, you can even perform this experiment blindfolded! Try to follow the line by just listening to the instructions coming from your RCX, and compare the lines drawn by your pencil for different values of H.

Summary

Math is the kind of subject that people either love or hate. If you fall in the latter group, we can't blame you for having skipped most of the content of this chapter. Don't worry, there was nothing you can't live without, just make an effort to understand the part about multiplication and divisions, because if you ignore the possible side effects, you could end up with some bad surprises in your calculations.

Consider the other topics—averages, interpolation, hysteresis—to be like tools in your toolbox. Averages are a useful instrument to soften the differences between single readings and to ignore temporary peaks. They allow you to group a set of readings and consider it as a single value. When you are dealing with a flow of data coming from a sensor, the moving average is the right tool to process the last **n** readings. The larger **n** is, the more the smoothing effect on the data.

Weighted averages have an advantage over simple averages in that they can show the trend in the data: You can assign the weights to put more importance on more recent data. Exponential smoothing is a special case of weighted averages, the results of which are particularly convenient on the implementation side, because they allow you to write compact and efficient code.

The interpolation technique proves useful when you want to estimate the value of a quantity that falls between two known limits. We described linear interpolation, which corresponds to tracing a straight line across two points in a graph. You then can use that line to calculate any value in the interval.

Hysteresis, a concept borrowed from physics, will help you in reducing the number of corrections your robots have to make to keep within a required behavior. By adding some hysteresis to your algorithms, your robots will be less reactive to changes. Hysteresis can also increase the efficiency of your system.

It's not necessary that you remember all the equations, just what they're useful for! You can always refer back to this chapter when you find a problem that might benefit from these mathematical tools.

Chapter 13

Knowing Where You Are

Solutions in this chapter:

- Choosing Internal or External Guidance

- Looking for Landmarks:
  Absolute Positioning

- Measuring Movement:
  Relative Positioning

Introduction

After our first few months of experimenting with robotics using the MIND-STORMS kit, we began to wonder if there was a simple way to make our robot know where it was and where it was going—in other words, we wanted to create some kind of navigation system able to establish its position and direction. We started reading books and searching the Internet, and discovered that this is still one of most demanding tasks in robotics and that there really isn't any single or simple solution.

In this chapter, we will introduce you to concepts of navigation, which can get very complex. We will start describing how positioning methods can be categorized into two general classes: *absolute* and *relative* positioning, the first based on external reference points, and the latter on internal measurements. Then we will provide some examples for both the categories, showing solutions and tricks that suit the possibilities of the MINDSTORMS system. In discussing absolute positioning, we will introduce you to navigation on pads equipped with grids or gradients, and to the use of laser beams to locate your robot in a room. As for relative positioning, we will explain how to equip your robot for the proper measurements, and will provide the math to convert those measurements into coordinates.

Choosing Internal or External Guidance

As we mentioned, there is no single method for determining the position and orientation of a robot, but you can combine several different techniques to get useful and reliable results. All these techniques can be classified into two general categories: *absolute* and *relative* positioning methods. This classification refers to whether the robot looks to the surrounding environment for tracking progress, or just to its own course of movement.

Absolute positioning refers to the robot using some external reference point to figure out its own position. These can be landmarks in the environment, either natural landmarks recognized through some kind of artificial vision, or more often, artificial landmarks easily identified by your robot (such as colored tape on the floor). Another common approach includes using radio (or light) beacons as landmarks, like the systems used by planes and ships to find the route under any weather condition. Absolute positioning requires a lot of effort: You need a prepared environment, or some special equipment, or both.

Relative positioning, on the other hand, doesn't require the robot to know anything about the environment. It deduces its position from its previous (known) position and the movements it made since the last known position. This is usually achieved through the use of encoders that precisely monitor the turns of the wheels, but there are also inertial systems that measure changes in speed and direction. This method is also called *dead reckoning* (short for *deduced reckoning)*.

Relative positioning is quite simple to implement, and applies to our LEGO robots, too. Unfortunately, it has an intrinsic, unavoidable problem that makes it impossible to use by itself: It accumulates errors. Even if you put all possible care into calibrating your system, there will always be some very small difference due to slippage, load, or tire deformation that will introduce errors into your measurements. These errors accumulate very quickly, thus relegating the utility of relative positioning to very short movements. Imagine you have to measure the length of a table using a very short ruler: You have to put it down several times, every time starting from the point where you ended the previous measurement. Every placement of the ruler introduces a small error, and the final result is usually very different from the real length of the table.

The solution employed by ships and planes, which use beacons like Loran or Global Positioning Systems (GPS) systems, and more recently by the automotive industry, is to combine methods from the two groups: to use dead reckoning to continuously monitor movements and, from time to time, some kind of absolute positioning to zero the accumulated error and restart computations from a known location. This is essentially what human beings do: When you walk down a street while talking to a friend, you don't look around continuously to find reference points and evaluate your position; instead, you walk a few steps looking at your friend, then back to the street for an instant to get your bearings and make sure you haven't veered off course, then you look back to your friend again.

You're even able to safely move a few steps in a room with your eyes shut, because you can deduce your position from your last known one. But if you walk for more than a few steps without seeing or touching any familiar object, you will soon lose your orientation.

In the rest of the chapter, we will explore some methods for implementing absolute and relative positioning in LEGO robots. It's up to you to decide whether or not to use any one of them or a combination in your applications. Either way, you will discover that this undertaking is quite a challenge!

Looking for Landmarks: Absolute Positioning

The most convenient way to place artificial landmarks is to put them flat on the floor, since they won't obstruct the mobility of your robot and it can read them with a light sensor without any strong interference from ambient light. You can stick some self adhesive tape directly on the floor of your room, or use a sheet of cardboard or other material over which you make your robot navigate.

Line following, which we have talked a lot about, is probably the simplest example of navigation based on using an artificial landmark. In the case of line following, your robot knows nothing about where it is, because its knowledge is based solely on whether it is to the right or left of the line. But lines are indeed an effective system to steer a robot from one place to another. Feel free to experiment with line following; for example, create some interruptions in a straight line and see if you are able to program your robot to find the line again after the break. It isn't easy. When the line ends, a simple line follower would turn around and go back to the other side of the line. You have to make your software more sophisticated to detect the sudden change and, instead of applying a standard route correction, start a new searching algorithm that drives the robot toward a piece of line further on. Your robot will have to go forward for a specific distance (or time) corresponding to the approximate length of the break, then turn left and right a bit to find the line again and resume standard navigation.

When you're done and satisfied with the result, you can make the task even more challenging. Place a second line parallel to the first, with the same interruptions, and see if you can program the robot to turn 90 degrees, intercept the second line, and follow that one. If you succeed in the task, you're ready to navigate a grid of short segments, either following along the lines or crossing over them like a bar code.

You can improve your robot navigation capabilities, and reduce the complexity in the software, using more elaborate markers. As we explained in Chapter 4, the LEGO light sensor is not very good at distinguishing different colors, but is able to distinguish between differences in the intensity of the reflected light. You can play with black and gray tapes on a white pad, and use their color as a source of information for the robot. Remember that a reading at the border between black and white can return the same value of another on plain gray. Move and turn your robot a bit to decode the situation properly, or employ more than a single light sensor if you have them.

Instead of placing marks on the pad, you can also print on it with a special black and white gradient. For example, you can print a series of dots with an intensity proportional to their distance from a given point a. The closer to a, the darker the point; a is plain black (see Figure 13.1). On such a pad, your robot will be able to return to a from any point, by simply following the route until it reads the minimum intensity.

Figure 13.1 A Gradient Pad with a Single Attractor

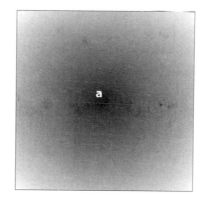

The same approach can be used with two points a and b, one being white and the other black. Searching for the whitest route, the robot arrives at a, while following the darkest it goes to b (Figure 13.2). We first saw this trick applied during the 1999 Mindfest gathering at the Massachusetts Institute of Technology (MIT): two robots were playing soccer, searching for a special infrared (IR) emitting ball. When one got the ball, it used the pad to find the proper white or black goal. Months later, we successfully replicated this setup with Marco Berti during a demonstration at an exhibition in Italy.

Figure 13.2 A Gradient Pad with Two Attractors

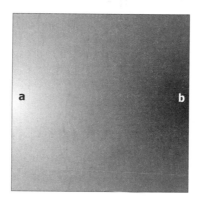

There are other possibilities. People have suggested using bar codes on the floor: When the robot finds one, it aligns and reads it, decoding its position from the value. Others tried complex grids made out of stripes of different colors. Unfortunately, there isn't a simple solution valid for all cases, and you will very likely be forced to use some dead reckoning after a landmark to improve the search.

Designing & Planning...

Making a Gradient Pad

To print a gradient pad with a single attractor A simply make the darkness (or brightness) of any point proportional to its distance from A. If *ax* and *ay* are the coordinates of A, and *x, y* are the coordinates of the given pixel, the distance will be:

```
dist = sqrt((x-ax)*(x-ax)+(y-ay)*(y-ay))
```

Scale the distance so as to have 100 percent black in A and 0 percent at the maximum distance from A measured in your pad, then multiply it by the constant that represents the maximum brightness of a pixel in your system:

```
intensity = dist/maxdist*maxbrite
```

Now apply this value to all three color components of a pixel using something similar to:

```
pixels[x,y] = rgb(intensity,intensity,intensity )
```

To generate a pad with two attractors A (*ax, ay*) and B (*bx, by*) respectively white and black, make the intensity of each pixel proportional to the ratio of the distance from B over the sum of the distances from A and from B:

```
adist = sqrt((x-ax)*(x-ax)+(y-ay)*(y-ay))
bdist = sqrt((x-bx)*(x-bx)+(y-by)*(y-by))
intensity = bdist/(adist+bdist)*maxbrite
pixels[x,y] = rgb(intensity,intensity,intensity)
```

Following the Beam

In the real world, most positioning systems rely on beacons of some kind, typically radio beacons. By using at least three beacons, you can determine your position on a two-dimensional plane, and with four or more beacons you can compute your position in a three-dimensional space. Generally speaking, there are two kinds of information a beacon can supply to a vehicle: its distance and its heading (direction of travel). Distances are computed using the amount of time that a radio pulse takes to go from the source to the receiver: the longer the delay, the larger the distance. This is the technique used in the Loran and the GPS systems. Figure 13.3 shows why two stations are not enough to determine position: There are always two locations A and B that have the same distance from the two sources.

Figure 13.3 There are Two Locations with the Same Distance from Two Stations

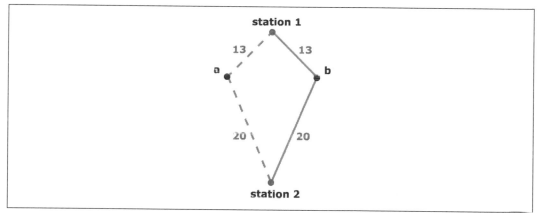

Adding a third station, the system can solve the ambiguity, provided that this third station does not lie along the line that connects the previous two stations (see Figure 13.4).

The stations of the VHF Omni-directional Range system (VOR) can not tell you the distance from the source of the beacon, but they do tell you their heading, that is, the direction of the route you should go to reach each station. Provided that you also know your heading to the North, two VOR stations are enough to locate your vehicle in most cases. Three of them are required to cover the case where the vehicle is positioned along the line that connects the stations, and as for the Loran and GPS systems, it's essential that the third station itself does not lay along that line (see Figure 13.5).

Figure 13.4 Three Nonaligned Stations Allow for Positioning with No Ambiguities Using Distances

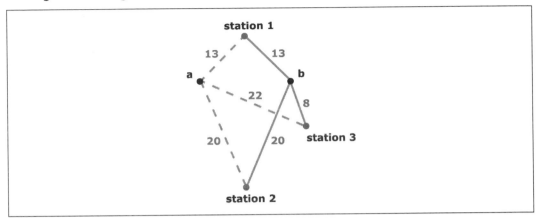

Figure 13.5 VOR-Like Systems Allow Positioning through the Headings of the Stations

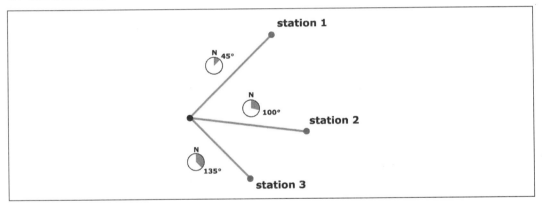

Using three stations, you can do without a compass, that is, you don't need to know your heading to the North. The method requires that you know only the angles between the stations as you see them from your position (see Figure 13.6).

To understand how the method works, you can perform a simple experiment. Take a sheet of paper and mark three points on it that correspond to the three stations. Now take a sheet of transparent material and put it over the previous sheet. Spot a point anywhere on it which represents your vehicle, and draw three lines from it to the stations, extending the lines over the stations themselves. Mark the lines with the name of the corresponding stations. Now, move the transparent sheet and try to intersect the lines again with the stations. There's an unlimited number of positions which connect two of the three stations, but there's only one location which connects all three of them.

Figure 13.6 Three Nonaligned Stations Allow for Positioning with No Ambiguities Using Angles

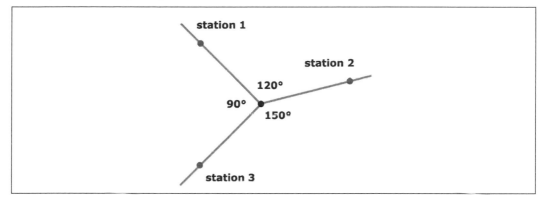

If you want to give this approach a try, the first problem you have to solve is that there's currently nothing in the LEGO world that can emit beacons of any kind, so you have to look for some alternative device. Unless you're an electrical engineer and are able to design and build your own custom radio system, you better stick with something simple and easy to find. The source need not be necessarily based on radio waves—light is effective as well, and we already have such a detector (the light sensor) ready to interface to the RCX.

By using light sources as small lighthouses, you can, in theory, make your robot find its way. But there are a few difficulties to overcome first:

- The light sensor isn't directional—you must shield it somehow to narrow its angle.

- Ambient light introduces interference, so you must operate in an almost lightless room.

- For the robot to be able to tell the difference between the beacons, you must customize each one; for example, making them blink at different rates (as real lighthouses do).

Laser light is probably a better choice. It travels with minimum diffusion, so when it hits the light sensor, it is read at almost 100 percent. Laser sources are now very common and very cheap. You can find small battery-powered pen laser pointers for just a few dollars.

If you have chosen laser as a source of light, you don't need to worry about ambient light interference. But how exactly would you use laser? Maybe by making some rotating laser lighthouses? Too complex. Let's see what happens if we revert the problem and put the laser source *on the robot*. Now you need just one laser, and can rotate it to hit the different stations. So, the next hurdle is to figure out how you know when you have hit one of those stations. If you place an RCX with a light sensor in every station, you can make it send back a message when it gets hit by the laser beam, and using different messages for every station, make your robot capable of distinguishing one from another.

When we actually tried this, we discovered that the light sensor is a very small target to hit with a laser beam, and as a result, realized we had set an almost impossible goal. To stick with the concept but make things easier, we discovered you could build a sort of diffuser in front of it to have a wider detection area. Jonathan Brown suggested one made with a simple piece of paper, which worked very well.

Now you have a working solution, but it's a pity you need so many RCXs, at least three for the stations and one for your robot. Isn't there a cheaper option? A different approach involves employing the simple plastic reflectors used on cars, bikes, and as cat's-eyes on the side of the road. They have the property of reflecting any incoming light precisely back in the direction from which it came. Using those as passive stations, when your robot hits them with its laser beam they reflect it back to the robot, where you have placed a light sensor to detect it.

This really seems the perfect solution, but it actually still has its weak spots. First, you have lost the ability to distinguish one station from the other. You also have to rely on dead reckoning to estimate the heading of each station. We explained that dead reckoning is not very accurate and tends to accumulate errors, but it can indeed provide you with a good *approximation* of the expected heading of each station, enough to allow you to distinguish between them. After having received the actual readings, you will adjust the estimated heading to the measured one. The second flaw to the solution is that most of the returning beam tends to go straight back to the laser beam. You must be able to very closely align the light sensor to the laser source to intercept the return beam, and even with that precaution, detecting the returning beam is not very easy.

To add to these difficulties, there is some math involved in deducing the position of the robot from the beacons, and it's the kind of math whose very name sends shivers down most students spines: trigonometry! (Don't worry, you can find formulas in some of the resources referenced by Appendix A.) This leads to another problem: The standard firmware has no support for trig functions, and though in theory you could implement them in the language Not Quite C (NQC) using some tables and interpolation, the LEGO firmware does not give you enough variables to get useful results. If you want to proceed with using beacons, you really have to switch to leJOS or legOS, which both provide much more computational power.

If you're not in the mood to face the complexity of trigonometry and alternative firmware, you can experiment with simpler projects that still involve laser pointers and reflectors. For example, you can make a robot capable of "going home." Place the reflector at the home base of the robot, that is, the location where you want it to return. Program the robot to turn in place with the laser active, until the light beam intercepts the reflector and the robot receives the light back, then go straight in that direction, checking the laser target from time to time to adjust the heading.

Measuring Movement: Relative Positioning

Relative positioning, or dead reckoning, is based on the measurement of either the movements made by the vehicle or the force involved (acceleration). We'll leave this latter category alone, as it requires expensive inertial sensors and gyroscopic compasses that are not easy to interface to the RCX!

The technique of measuring the movement of the vehicle, called *odometry*, requires an *encoder* that translates the turn of the wheels into the corresponding traveled distance. Choosing among LEGO supplies, the obvious candidate is the rotation sensor. However, you already know from Chapter 4 that you can emulate the rotation sensor with a touch or a light sensor. You typically will need two of them, though by using some of the special platforms seen in Chapter 8 (the dual differential drive and synchro drive) you can implement odometry with a single rotation sensor. If you don't have any, look in Chapter 4 for some possible substitutes.

The equations for computing the position from the decoded movements depends on the architecture of the robot. We will explain it here using the example of the differential drive, once again referring you to Appendix A for further resources on the math used.

Suppose that your robot has two rotation sensors, each connected through gearing to one of the main wheels. Given D as the diameter of the wheel, R as the resolution of the encoder (the number of counts per turn), and G the gear ratio between the encoder and the wheel, you can obtain the conversion factor F that translates each unit from the rotation sensor into the corresponding traveled distance:

$$F = (D \times \pi) \; / \; (G \times R)$$

The numerator of the ratio, $D \times \pi$, expresses the circumference of the wheel, which corresponds to the distance that the wheel covers at each turn. The denominator of the ratio, $G \times R$, defines the increment in the count of the encoder (number of *ticks*) that corresponds to a turn of the wheel. F results in the unit distance covered for every tick.

Your robot uses the largest spoked wheels, which are 81.6mm in diameter. The rotation sensor has a resolution of 16 ticks per turn, and it is connected to the wheel with a 1:5 ratio (five turns of the sensor for one turn of the wheel). The resulting factor is:

$$F = 81.6 \text{ mm} \times 3.1416 \, / \, (5 \times 16 \text{ ticks}) \approx 3.2 \text{ mm/tick}$$

This means that every time the sensor counts one unit, the wheel has traveled 3.2mm. In any given interval of time, the distance T_L covered by the left wheel will correspond to the increment in the rotation sensor count I_L multiplied by the factor F:

$$T_L = I_L \times F$$

And similarly, for the right wheel:

$$T_R = I_R \times F$$

The centerpoint of the robot, the one that's in the middle of the ideal line that connects the drive wheels, has covered the distance T_C:

$$T_C = (T_R + T_L) \, / \, 2$$

To compute the change of orientation ΔO you need to know another parameter of your robot, the distance between the wheels B, or to be more precise, between the two points of the wheels that touch the ground.

$$\Delta O = (T_R - T_L) \, / \, B$$

This formula returns ΔO in radians. You can convert radians to degrees using the relationship:

$\Delta O_{Degrees} = \Delta O_{Radians} \times 180 / \pi$

You can now calculate the new relative heading of the robot, the new orientation O at time i based on previous orientation at time i − 1 and change of orientation ΔO. O is the direction your robot is pointed at, and results in the same unit (radians or degrees) you choose for ΔO.

$O_i = O_{i-1} + \Delta O$

Similarly, the new Cartesian coordinates of the centerpoint come from the previous ones incremented by the traveled distance:

$x_i = x_{i-1} + T_C \times \cos O_i$
$y_i = y_{i-1} + T_C \times \sin O_i$

The two trigonometric functions convert the vectored representation of the traveled distance into its Cartesian components.

O' this villainous trigonometry again! Unfortunately, you can't get rid of it when working with positioning. Thankfully, there are some special cases where you can avoid trig functions, however; for example, when you're able to make your robot turn in place precisely 90 degrees, and truly go straight when you expect it to. In this situation, either x or y remains constant, as well as the other increments (or decrements) of the traveled distance T_C. Thinking back to Chapter 8, two platforms make ideal candidates for this: the dual differential drive and the synchro drive.

Using a dual differential drive you need just one rotation sensor, attached to either the right or the left wheel. The mechanics guarantees that when one motor is on, the robot drives perfectly straight, while when the other is on, the robot turns in place. In the first case, the rotation sensor will measure the traveled distance T_C, while in the second, you must count its increments to turn exactly in multiples of 90 degrees. In Chapter 23, you will find a robot which turns this theory into a practical application, a dual differential drive that navigates using a single rotation sensor. We will go through the math again and will describe a complete NQC implementation of the formulas.

The synchro drive can be easily limited to turn its wheels only 90 degrees. With this physical constraint regarding change of direction, you can be sure it will proceed using only right angles. Connect the rotation sensor to the motion motor. Just as in the previous case you will use it to measure the traveled distance. In this setup, as with that using the dual differential drive, one rotation sensor is enough.

Summary

We promised in the introduction that this was a difficult topic, and it was. Nevertheless, making a robot that has even a rough estimate of its position is a rewarding experience.

There are two categories of methods for estimating position, one based on absolute positioning and the other on relative positioning. The first category usually requires artificial landmarks or beacons as external references, both of which we explored. Artificial landmarks can range from a simple line to follow, to a grid of marks, to more complex gradient pads. All of them allow a good control of navigation through the use of a light sensor. The MINDSTORMS line is not very well equipped for absolute positioning through the use of beacons; however, we described some possible approaches based on laser beams and introduced you to the difficulties that they entail.

Relative positioning is more easily portable to LEGO robots. We explained the math required to implement deduced reckoning in a differential drive robot, and suggested some alternative architectures that help in simplifying the involved calculations. Unfortunately, relative positioning methods have an unavoidable tendency to accumulate errors; you can reduce the magnitude of the problem, but you cannot eliminate it. This is the reason that real life navigation systems—like those used in cars, planes, and ships—usually rely on a combination of methods taken from both categories: Dead reckoning isn't very computation intensive and provides good precision in a short range of distances, while absolute references are used to zero the errors it accumulates.

Part II

Projects

Chapter 14

Classic Projects

Solutions in this chapter:

- **Exploring Your Room**
- **Following a Line**
- **Modeling Cars**

Introduction

From this chapter on, we will explore several example projects that could be the inspiration for many others of your own creation. As we already explained, the spirit of the book is not to provide you with step-by-step instructions, but rather to give you a foundation of information and let your imagination and creativity do the rest. For this reason, you will find some pictures of each model, some text that describes their distinguishing characteristics, and tricks that could be useful for other projects. Of course, we don't expect each detail to be visible in the pictures. It isn't important that your models look exactly like ours!

Another point we want to bring to your attention is that there is no reason to read the project chapters in Part II in order. Feel free to jump to the project that attracts you most, since they aren't ordered according to their level of difficulty.

In this chapter, we'll show you some projects that could be considered "classic," because almost everybody with a MINDSTORMS kit tries them sooner or later. Though you might not find them exciting, working with them is a good way to build up some solid experience and learn tricks that will prove useful in more complex projects. If this is among your first forays into robotics, we strongly suggest you dedicate some time to them.

All the robots appearing in this chapter have been built solely from the RIS 1.5 equipment. Only in describing some of the possible additions to the robots do we suggest extra parts.

Exploring Your Room

Well, actually "exploring" your room is too strong a term for what we are proposing here, it's more like *surviving* your room—your robot and your furniture could take some hits! The task here is to build a robot with the basic ability to move around, detect obstacles, and change its route accordingly.

For simplicity of design, and for the robot's ability to turn in place, we suggest you make this robot from a differential drive architecture, like the one shown in Figure 14.1.

We deliberately chose a gear ratio that makes the robot rather slow: 1:9, obtained from two 1:3 stages (Figure 14.2). This ensures that if you make some error in the code and the robot fails to properly detect the obstacle, it won't collide with it at too high a speed. Never expect everything to go well on the first try—because it won't!

Figure 14.1 Start with a Simple Differential Drive

Figure 14.2 Detail of the Two-Stage Gear Train

When you feel satisfied with your software and your robot runs safely around your room, you can always try a faster ratio. Substituting the second 1:3 gearing with two 16t gears will give you an overall 1:3 ratio, making your robot about three times faster.

This robot has been designed to host the RCX behind the motors, and without the weight of the RCX in that position it actually does a wheelie (flips up)!

Another thing we'd like you to notice is that we made the robot rather large, keeping the main wheels far from the body of the robot. There's a reason for this, too: In a differential drive, the distance between the drive wheels affects the turning speed of the robot, because the wheels have to cover a longer distance during turns. The further the wheels are from the midpoint, the slower the turns. Since you're going to control turns through timing, slow turns are a desirable property which means finer movement control.

The caster wheel is the same kind we showed in Chapter 8. Now add the RCX and a couple of bumpers that are normally closed, like those in Chapter 4 (see Figure 14.3), and you're ready to go—well, ready to program the robot, anyway. Check out Figure 14.4 to see what the completed robot looks like.

Figure 14.3 Detail of the Bumper

Figure 14.4 The Robot, Complete with RCX and Bumpers

The program itself is very simple: Go straight until one of the touch sensors opens. When that happens, reverse for a few fractions of a second, then turn in place, right or left depending on which bumper found the obstacle. Finally, resume straight motion.

Experiment with different timing for turns, until you are happy with the result. You might also use some random values for turns to make the behavior of your robot a bit less predictable and thus more interesting. If you feel at ease with the programming, you can add more intelligence to your creature—for example, to make it capable of realizing when it's stuck in a cul-de-sac. This can be achieved by monitoring the number of collisions in a given time, or the average time elapsed between the last *n* collisions, and then adopting a more radical behavior (like turning 180°).

Detecting Edges

If your room has a flight of stairs going down, you can equip the robot with a kind of detector to sense the edge and avoid a bad fall. Normally, you would use a touch sensor for this, connecting it to a feeler flush with the ground (see Figure 14.5 for a detail). When the feeler in front of the robot drops, you have detected an edge.

Unless you have a third touch sensor, you are forced to use the light sensor. It's time to look back at some of the tricks explained in Chapter 4 and see if you find something useful. A light sensor can actually emulate a touch sensor: You have to place movable parts of different colors in front of it, so that when contact is made, the parts move, and the color of the brick in front of the light sensor changes.

Figure 14.5 Edge Detection System Detail

We kept the edge sensor behind the bumpers, so that in most cases it doesn't interfere in the obstacle detection (Figure 14.6).

Unfortunately, this system doesn't cover all possible scenarios, because your robot could approach the edge at an angle that allows a wheel over the edge before detection occurs. You can improve upon the design and avoid this by providing the robot with two left- and right-edge sensors, but you'll probably have to give up the double bumper and go with a single sensor bumper.

Figure 14.6 The Edge Detection System Installed on the Robot

Using a different approach, you could write the software to make the robot very cautious, turning slightly left and right from time to time to see if there's a dangerous precipice around.

Variations on Obstacle Detection

If you own a couple of rotation sensors, you can experiment with indirect obstacle detection. Connect them to the main wheels, and program the robot to monitor their count while in motion. If both the motors are on forward, but the count doesn't increase, the robot knows an obstacle has blocked it. As a positive side effect, the rotation sensors allow you to use the same platform for experimenting with navigation, applying some of the concepts about dead reckoning explained in Chapter 13.

You can also implement indirect obstacle detection using a *drag sensor*. The idea requires that your robot keep a mobile part in touch with the ground, and that the friction that this part exerts against the floor surface when the robot moves activates a touch sensor. For example, you can use the friction of a rubber tire to oppose the force of a rubber band that keeps a touch sensor closed. When the robot moves, the friction of the tire on the floor overcomes the force of the rubber band and opens the touch sensor; as soon as the robot stops—or has gotten blocked by an obstacle—the friction disappears and the touch sensor closes.

Following a Line

The line-following theme is often mentioned in Part I of the book, as we think it is a very useful indicator of how different techniques can improve the behavior of a robot. The time has come to give it an official place, and face the topic in its entirety. Let's review what we have already said about line following:

- You must actually follow the edge between the tape and the floor, reading an average value between dark and bright, so when you read too dark or too bright you know which direction to turn to find the route back (Chapter 4).

- If you want to keep your software and robot as general as possible, you should use some kind of self-calibration process before the actual following begins. Calibration consists of taking readings of light and dark areas on the pad before actually starting line following. This lets your robot adjust its parameters to the actual lighting conditions at the time it runs, which are almost certainly different from the conditions it was designed in (Chapter 6).

- Some platforms can benefit from the introduction of a small quantity of hysteresis to reduce the number of corrections and get a higher efficiency. We explained in Chapter 12 that hysteresis widens the gray area between light and dark, which keeps the robot from spending too much time on course correction instead of moving forward!

To turn theory into practice and experiment with line following, you can use the same differential drive as in the previous project in this chapter. Remove the bumpers and mount a light sensor facing down, as shown in Figure 14.7.

The light sensor is stacked on a 2 x 4 black brick, which is attached to the structure with two black pegs placed into its tubes at the bottom. In fact, the diameter of those tubes is very close to one of the holes in the beams, so the pegs and axles fit very well (isn't LEGO wonderful?) The distance of the sensor from the pad is very important (Figure 14.8). Our experience teaches us that for the best results this distance should fall in the range of 1mm to 5mm (0.04 to 0.2 inches).